Lecture Notes in Computer Science

Lecture Notes in Artificial Intelligence   14456

Founding Editor

Jörg Siekmann

Series Editors

Randy Goebel, *University of Alberta, Edmonton, Canada*
Wolfgang Wahlster, *DFKI, Berlin, Germany*
Zhi-Hua Zhou, *Nanjing University, Nanjing, China*

The series Lecture Notes in Artificial Intelligence (LNAI) was established in 1988 as a topical subseries of LNCS devoted to artificial intelligence.

The series publishes state-of-the-art research results at a high level. As with the LNCS mother series, the mission of the series is to serve the international R & D community by providing an invaluable service, mainly focused on the publication of conference and workshop proceedings and postproceedings.

Francesco Amigoni · Arunesh Sinha
Editors

Autonomous Agents and Multiagent Systems

Best and Visionary Papers

AAMAS 2023 Workshops
London, UK, May 29 – June 2, 2023
Revised Selected Papers

Editors
Francesco Amigoni
Politecnico di Milano
Milan, Italy

Arunesh Sinha
Rutgers University
Piscataway, NJ, USA

ISSN 0302-9743   ISSN 1611-3349 (electronic)
Lecture Notes in Artificial Intelligence
ISBN 978-3-031-56254-9   ISBN 978-3-031-56255-6 (eBook)
https://doi.org/10.1007/978-3-031-56255-6

LNCS Sublibrary: SL7 – Artificial Intelligence

This Springer imprint is published by the registered company Springer Nature Switzerland AG
The registered company address is: Gewerbestrasse 11, 6330 Cham, Switzerland

Paper in this product is recyclable.

Preface

AAMAS (the International Conference on Autonomous Agents and Multiagent Systems) is the largest and most influential conference in the area of agents and multiagent systems. It is organized annually by the International Foundation for Autonomous Agents and Multiagent Systems (IFAAMAS). The AAMAS conference series was initiated in 2002 in Bologna, Italy, as a joint event comprising the 6th International Conference on Autonomous Agents (AA), the 5th International Conference on Multiagent Systems (ICMAS), and the 9th International Workshop on Agent Theories, Architectures, and Languages (ATAL). In 2023, the 22nd edition of AAMAS was held in London, after three editions were held online due to the restrictions imposed by the COVID-19 pandemic.

Besides the main program, AAMAS 2023 hosted a rich workshop program, aimed at stimulating and facilitating discussion, interaction, and comparison of approaches, methods, and ideas related to specific topics, both theoretical and applied, in the general area of autonomous agents and multiagent systems. Workshops provide an informal setting where participants have the opportunity to discuss specific technical topics in an atmosphere that fosters the active exchange of ideas and supports community development. In 2023, AAMAS hosted a total of 14 workshops.

This volume contains a selection of papers from 7 of these workshops:

- Workshop on Autonomous Robots and Multirobot Systems (ARMS)
- Workshop on Adaptive and Learning Agents (ALA)
- Workshop on Interdisciplinary Design of Emotion Sensitive Agents (IDEA)
- Workshop on Rebellion and Disobedience in Artificial Intelligence (RaD-AI)
- Workshop on Neuro-symbolic AI for Agent and Multi-Agent Systems (NeSyMAS)
- Workshop on Multiagent Sequential Decision Making under Uncertainty (MSDM)
- Workshop on Citizen-Centric Multi-Agent Systems (C-MAS)

In particular, we include 5 best papers from ALA, IDEA, NeSyMAS, MSDM, and C-MAS workshops and 7 papers deemed as visionary, one for each of the above workshops. Given the specificities and scope of the different workshops, the nomination of both the best and the most visionary papers of each workshop relied on the corresponding organizing committee, which was charged with defining the criteria for selection and nominating the papers to be included in this volume. The criteria included the relevance, originality, novelty, significance, and technical quality of the papers and sometimes took into account the reviewers' comments and recommendations.

As workshop chairs, we believe that the AAMAS workshops provide a unique opportunity for researchers to connect and exchange ideas with other researchers working on very closely related topics and problems. We hope that the publication of this volume can help to disseminate the high-quality work that was presented and discussed in these workshops and serve as an instrument to foster further participation in the AAMAS workshop program.

December 2023   Francesco Amigoni
                                                      Arunesh Sinha

Organization

AAMAS 2023 Workshop Co-chairs

Francesco Amigoni   Politecnico di Milano, Italy
Arunesh Sinha   Rutgers University, USA

AAMAS 2023 Workshop Organizers

ARMS 2023

Nicola Basilico   University of Milan, Italy
Mohan Sridharan   University of Birmingham, UK

ALA 2023

Francisco Cruz   UNSW Sydney, Australia
Conor F. Hayes   Lawrence Livermore National Laboratory, USA
Caroline Wang   University of Texas at Austin, USA
Connor Yates   Oregon State University, USA

IDEA 2023

Loïs Vanhée   Umeå University, Sweden
Carole Adam   Université Grenoble Alpes, France
Melania Borit   UiT The Arctic University of Norway, Norway
Christopher Gagne   Max Planck Institute for Biological Cybernetics, Germany

RaD-AI 2023

David Aha   Naval Research Laboratory, USA
Gordon Briggs   Naval Research Laboratory, USA
Reuth Mirsky   Bar-Ilan University, Israel
Ram Rachum   Bar-Ilan University, Israel

Kantwon Rogers   Georgia Tech, USA
Peter Stone   University of Texas at Austin, USA and
                                   Sony AI, USA

NeSyMAS 2023

Vaishak Belle   University of Edinburgh, UK
Michael Fisher   University of Manchester, UK
Xiaowei Huang   University of Liverpool, UK
Masoumeh Mansouri   University of Birmingham, UK
Albert Meroño-Peñuela   King's College London, UK
Sriraam Natarajan   UT Dallas, USA
Efi Tsamoura   Samsung Cambridge, UK

MSDM 2023

Yifeng Zeng   Northumbria University, UK
Yuchen Xiao   J.P. Morgan, USA
Yinghui Pan   Shenzhen University, China
Prashant Doshi   University of Georgia, USA

C-MAS 2023

Behrad Koohy   University of Southampton, UK
Kate Larson   University of Waterloo, Canada
Marija Slavkovik   University of Bergen, Norway
Natalia Criado   Universitat Politècnica de València, Spain
Sebastian Stein   University of Southampton, UK
Vahid Yazdanpanah   University of Southampton, UK

Contents

Best Papers

Fair Deep Reinforcement Learning with Generalized Gini Welfare Functions

Guanbao Yu[1], Umer Siddique[2], and Paul Weng[1(✉)] [iD]

[1] UM-SJTU Joint Institute, Shanghai Jiao Tong University, Shanghai, China
{gbyu66,paul.weng}@sjtu.edu.cn
[2] University of Texas at San Antonio, San Antonio, Texas, USA

Abstract. Learning fair policies in reinforcement learning (RL) is important when the RL agent's actions may impact many users. In this paper, we investigate a generalization of this problem where equity is still desired, but some users may be entitled to preferential treatment. We formalize this more sophisticated fair optimization problem in deep RL, provide some theoretical discussion of its difficulties, and explain how existing deep RL algorithms can be adapted to tackle it. Our algorithmic innovations notably include a state-augmented DQN-based method for learning stochastic policies, which also applies to the usual fair optimization setting without any preferential treatment. We empirically validate our propositions and analyze the experimental results on several application domains.

Keywords: Deep reinforcement learning · Fair Optimization · Multi-objective

1 Introduction

Deep reinforcement learning (RL) has attained considerable attention in recent years, presenting promising avenues for the development of adaptive learning agents [16,23,27,28]. These agents, when deployed in real applications, such as traffic light control, software-defined networking, or data centers, interact with and impact many users. Hence, for these systems to be accepted by end-users when they are in operation, fairness needs to be taken into account in their design.

Fairness is rooted in the principle of "equal treatment of equals", which, in simple terms, states that individuals with similar characteristics should be treated similarly. Previous work [10,46] in learning fair policies in RL has primarily focused on this principle, under the assumption that all individuals are equal, which may not be suitable for all applications. For instance, in various contexts such as software-defined networking and data centers, service providers customarily offer different levels of Quality of Service (QoS) to different user tiers. In such cases, although the principle of "equal treatment of equals" is still a desired objective, higher-paying users should arguably be entitled to higher priority or better services.

F. Amigoni and A. Sinha (Eds.): AAMAS 2023 Workshops, LNAI 14456, pp. 3–29, 2024.
https://doi.org/10.1007/978-3-031-56255-6_1

In this work, we deliberately relax the assumption of equal individuals and consider the more general case where different users may have different rights. Our goal is to investigate this more sophisticated fairness problem in the context of deep RL. Here, we aspire to learn policies that ensure that while certain users may receive preferential treatment, those with similar rights are fairly treated.

Contributions. We formalize this novel problem in deep RL as a fair optimization problem (Sect. 4.2). We discuss the theoretical aspects and difficulties of this problem (Sect. 4.3). Based on this discussion, we propose several adaptations of deep RL algorithms to solve this problem (Sect. 5). Notably, we design a novel state-augmented DQN-based method for learning fair stochastic policies. Finally, we experimentally validate our propositions (Sect. 6).

2   Related Work

Recent work in fairness studies has introduced various notions of fairness, encompassing both group fairness and individual fairness. While the majority of the literature in this line of research focuses on the foundational principle of ensuring equitable treatment for all users, regardless of whether they are considered as groups or individuals, our work takes a different approach. It essentially aligns with the individual fairness category and introduces the notion of providing preferential treatment to specific users while simultaneously ensuring equitable treatment for all other users.

2.1   Fairness in AI and ML

Due to the realization of the tremendous impact that artificial intelligence (AI) and machine techniques (ML) can have on our lives, fairness has recently become an important and active research direction   [1,7,12,18,30,44,47,53,55,56]. Although various definitions of fairness have been considered in AI, e.g., proportionality [3,50] or envy-freeness [11] and its multiple variants (e.g., [4,9]), the majority of this literature in machine learning focuses on the impartiality aspect of fairness: "equal treatment of equals". Proposed methods in this direction typically rely on a constraint-based or penalty-based formulation in order to control bias at the individual or group level. In contrast, our work is based on studies in distributive justice [5,29,41]. We aim at optimizing a social welfare function that encodes impartiality, but also equity and efficiency (see Sect. 3.4 for more details). This principled approach has also been recently advocated in several recent papers [15,20,49,54] and applied in various machine learning tasks, such as sequential decision-making, which we discuss below, but also ranking [17] for instance.

In mathematical optimization, such an approach is called fair optimization [35]. Many continuous and combinatorial optimization problems in various application domains [2,33,34,36,45] have been extended to optimize for fairness.

In this direction, the closest work [36] regards fair optimization in Markov decision processes. However, the methods proposed in this direction typically assume that the model is known and therefore, they do not require learning.

2.2 Fairness in RL

Fairness in RL starts to receive more attention. Different directions have been studied, e.g., fairness constraint to reduce discrimination [53], fairness with respect to state visitation [19,21], the usual case of fairness with respect to agents [22], or the more general case of fairness with respect to users [10,26,46, 58]. For example, the work of Wen et al. [53] explored fairness constraints as a means to reduce discrimination, while Zimmer et al. [58] proposed a self-oriented team-oriented decentralized cooperative multi-agent framework (SOTO) that optimizes fairness and efficiency. Siddique et al. [46] investigated the problem of learning fair policies in RL by employing social welfare functions to ensure all users are treated fairly. In contrast, we extend this principled approach to a more general setting, relaxing the assumption that all users are equal.

2.3 State Augmentation

State augmentation (used in our DQN variants) has been exploited in various previous work, e.g., in MDPs [24] or more recently in safe RL [48], risk-sensitive RL [14], RL with delays [32], and partially-observable path planning [31]. However, to the best of our knowledge, this technique has not been applied in fair optimization. Moreover, our technique to learn stochastic policies in DQN is also novel.

3 Background

We first recall the Markov decision process (MDP) model and RL, then present the multi-objective extension of MDP. We also provide an overview of several deep RL algorithms. Finally, we review the social welfare functions (SWFs) that we used to encode fairness in deep RL.

Notations. Both the matrices and vectors are written in bold. For any vector $u \in \mathbb{R}^D$, $u^\uparrow$ corresponds to the vector with the components of vector u sorted in an increasing order (i.e., $u_1^\uparrow \le \ldots \le u_D^\uparrow$). For any integer $D > 0$, the $D - 1$ simplex is denoted by $\Delta_D = \{w \in \mathbb{R}^D \mid \sum_i w_i = 1 \text{ and } w_i \ge 0, i = 1, \ldots, D\}$. We denote $\mathbb{S}_D$ the symmetric group of degree D (i.e., set of permutations over $\{1, \ldots, D\}$). For any permutation $\sigma \in \mathbb{S}_D$ and vector $u \in \mathbb{R}^D$, vector u_σ denotes $(u_{\sigma(1)}, \ldots, u_{\sigma(D)})$.

3.1 Markov Decision Process and RL

A Markov Decision Process (MDP) [40] model is characterized by its set of states $\mathcal{S}$, set of actions $\mathcal{A}$, a transition model P which specifies the probability of reaching next state s' by taking action a in state s, and a reward function R indicating the immediate reward of performing action a in state s. This model also includes the discount factor $\gamma \in [0, 1)$, and the probability distribution over the initial states d_0.

A *policy* π in an MDP model provides guidance on which action to take in any state s. It can be deterministic if $\pi(s) = a$ or stochastic if $\pi(a \mid s) = Pr(a \mid s)$. Note that deterministic policies are special cases of stochastic ones. For a policy π, we denote P_π (resp. r_π) the transition (resp. reward) function induced by π, i.e., $P_\pi(s, s') = P(s, \pi(s), s')$ (resp. $r_\pi(s) = r_\pi(s, \pi(s))$). The usual goal in MDP is to learn a policy π that maximizes the expected discounted reward, i.e., $\mathbb{E}\left[\sum_{t=1}^{\infty} \gamma^{t-1} r_t\right]$.

Formally, the *(state) value function* $v_\pi : \mathcal{S} \to \mathbb{R}$ of a policy π from an initial state s is defined by:

$$v_\pi(s) = \mathbb{E}_{P,\pi}\left[\sum_{t=1}^{\infty} \gamma^{t-1} r_t \mid s\right], \tag{1}$$

where $\mathbb{E}_{P,\pi}$ is the expectation taken with respect to transition function P and policy π, and r_t is the random variable that represents the reward obtained at time step t. The value function v_π provides the expected discounted reward one can get by following the corresponding policy π from state s. Similarly, the *action-value function* $Q_\pi : \mathcal{S} \times \mathcal{A} \to \mathbb{R}$ is given by:

$$Q_\pi(s, a) = \mathbb{E}_{P,\pi}\left[\sum_{t=1}^{\infty} \gamma^{t-1} r_t \mid s, a\right]. \tag{2}$$

Formally, both the MDP and RL attempt to address the following optimization problem: $\operatorname{argmax}_\pi \sum_{s \in \mathcal{S}} d_0(s) v_\pi(s)$, where d_0 is the initial state distribution and v_π is the value function approximated by following the current policy π. A solution to this problem is an *optimal* policy, which is denoted by π^*.

3.2 Multiobjective Markov Decision Process

We formulate the novel fair optimization problem as a multiobjective MDP (MOMDP), where each objective corresponds to the individual utility of a user in our setting. Therefore, the rewards in MOMDPs are vectors instead of scalars. The reward function of a MOMDP can be formalized as $\boldsymbol{r}(s, a) \in \mathbb{R}^D$ where D is the number of objectives (users).

All the previous definitions in MDP can be naturally extended to MOMDP. For example, the value function in (1) now becomes:

$$\boldsymbol{v}_\pi(s) = \mathbb{E}_{P,\pi}\left[\sum_{t=1}^{\infty} \gamma^{t-1} \boldsymbol{r}_t \mid s\right], \tag{3}$$

where $r_t \in \mathbb{R}^D$ is the vector reward obtained at time step t and all the operations (addition, product) are component-wise. Similarly, the Q-value, as previously defined in (2), is redefined as:

$$Q_\pi(s, a) = \mathbb{E}_{P,\pi}\left[\sum_{t=1}^{\infty} \gamma^{t-1} r_t \mid s, a\right]. \tag{4}$$

Formally, solving an MOMDP amounts to solving the following multiobjective optimization problem: $\mathrm{argmax}_\pi \sum_{s \in S} d_0(s) v_\pi(s)$, where d_0 is the initial state distribution, v_π is the multiobjective version of value function that is approximated by following the current policy π and the vector maximization is with respect to *Pareto dominance*. Recall, *Pareto dominance* formally states: $\forall v, v' \in \mathbb{R}^D$, v *weakly Pareto-dominates* $v' \Leftrightarrow \forall i, v_i \geq v'_i$. Besides, v *Pareto-dominates* $v' \Leftrightarrow \forall i, v_i \geq v'_i$ and $\exists j, v_j > v'_j$. As there is no risk of confusion, Pareto dominance is simply denoted $\geq$ for its weak form and $>$ for its strict form.

Traditionally, in MOMDPs the goal is to find all *Pareto non-dominated* solutions. While this approach may work for small problems [52], the scalability concern becomes apparent when tackling large-scale decision-making systems. In such cases, the number of *Pareto non-dominated* solutions may grow exponentially with the size of the problem [39]. Moreover, in RL, the agent has to make a single decision at a given time step. To address the scalability challenge and enable the learning of a single, effective decision, we employ aggregation methods. These methods effectively transform the multiobjective optimization problem into a single objective formulation. Since our goal is to find balanced solutions, which amounts to finding fair solutions, we employed the type of aggregate methods that are socially fair. We explain those methods in Sect. 3.4.

3.3 Deep RL

Deep RL is the study of RL using neural networks as function approximators. They are needed to tackle large-scale RL problems, where the state and/or action spaces become large or continuous. With parametric function approximation such as neural networks, a function f is approximated by $\hat{f}_\theta$ where θ denotes the parameters of the parametric function, which can be learned during training. In RL, both value functions or policies can be approximated.

Deep Q-Network (DQN) [28] is an example of deep RL algorithm where the optimal Q function is approximated by a neural network with parameter θ. This Q-network takes a state s as input and outputs an estimated $\hat{Q}_\theta(s, a)$ for all actions. It is trained to minimize the following L_2 loss for a sampled transition (s, a, r, s'):

$$\left(r + \gamma \hat{Q}_{\theta'}(s', a^*) - \hat{Q}_\theta(s, a)\right)^2$$

where $a^* = \mathrm{argmax}_{a' \in \mathcal{A}}\left(r + \gamma \hat{Q}_{\theta'}(s', a')\right)$ and θ' represents the parameters of the target Q-network which promotes more stable training. The transitions

(s, a, r, s') are sampled from a replay buffer storing experiences generated from online interactions with the environment. The term $r + \gamma \hat{Q}_{\theta'}(s', a^*)$ is called *target Q-value*.

Policy gradient methods constitute another approach for solving RL problems. In contrast to value-based methods like DQN, policy gradient methods explicitly optimize the desired objective function in a parameterized policy space, with the goal of finding a policy $\pi_\theta(a \mid s)$ (θ being the policy parameters) that maximizes the expected sum of reward. In policy gradient methods, the objective function $J(\theta)$ can be formally defined as:

$$J(\theta) = \sum_{s \in S} d_{\pi_\theta}(s) V_{\pi_\theta}(s) = \sum_{s \in S} d_{\pi_\theta}(s) \sum_{a \in A} \pi_\theta(a \mid s) Q_{\pi_\theta}(s, a), \tag{5}$$

where $d_{\pi_\theta}(s)$ is the stationary state distribution under policy π_θ.

Parameter θ can be learned using gradient ascent by following the update direction given by the *Policy Gradient Theorem* [51]:

$$\nabla_\theta J(\theta) = \mathbb{E}_{s \sim d_\pi, a \sim \pi_\theta(\cdot \mid s)} [Q_{\pi_\theta}(s, a) \nabla_\theta \log \pi_\theta(a \mid s)]. \tag{6}$$

where the Q-value function $Q_{\pi_\theta}(s, a)$ can be estimated using Monte-Carlo or temporal difference methods.

3.4   Fairness

In this paper, an optimal fair solution is required to satisfy three properties [46]: efficiency, equity, and impartiality. The efficiency property states that a solution should be Pareto-optimal. This is a natural property because selecting a Pareto-dominated solution would be irrational. The equity property is based on the *Pigou-Dalton principle* [29], which states that transferring utility from a better-off user to a worse-off user results in a fairer solution. This principle establishes the foundation of fairness by distributing equal wealth among different users, which is a critical component in our definition of fairness. The impartiality property corresponds to the *"equal treatment of equals"* principle. This principle served as the foundation for previous works that assume all users are equal. However, in our work, we relax this assumption and consider a more general case in which some users may be given preference over others.

We rely on *social welfare functions (SWFs)* to formalize these three properties as an objective function. An SWF evaluates how good a solution is for all users by aggregating all users' utilities. In this paper, we only discuss those SWFs that satisfy our notion of fairness and refer to them as *fair SWFs*. One notable group of fair SWFs in the literature is the *generalized Gini social welfare function* (GGF), which is defined as follows:

$$GGF_{\boldsymbol{w}}(\boldsymbol{u}) = \sum_{i=1}^{D} \boldsymbol{w}_i \boldsymbol{u}_i^\uparrow, \tag{7}$$

where $u \in \mathbb{R}^D$ and $w \in \Delta_D$ is a fixed positive weight vector whose components are strictly decreasing (i.e., $w_1 > \ldots > w_D > 0$). Intuitively, by assigning larger weights on smaller utility values, GGF will yield larger scores when the utility distribution becomes more balanced while keeping the total utility constant.

GGF satisfies all three of the above-mentioned properties. Since GGF is a strictly increasing function with positive weights, it implies that it is monotonic in terms of Pareto-dominance and thus meets the efficiency property. GGF also satisfies the equity property because it is a strictly Schur-concave function, which implies that it is monotonic with respect to Pigou-Dalton transfers. Finally, because the components of GGF are symmetric (i.e., independent of the order of their arguments), it satisfies the impartiality property.

Despite the fact that GGF is a simple yet effective SWF for encoding fairness, it has some limitations. For instance, the symmetry of GGF entails that it only applies to cases where all users are equal. However, in many real-world applications, some objectives/users may be preferred. For instance, as discussed in the introduction, the service providers controlled by autonomous systems have to take different user tiers into account. For such systems, a fair SWF that can encode preferences over objectives is required, which we will explain in the following section.

4   Fair Optimization with Preferential Treatment

In this section, we first extend GGF to a generalized fair SWF that can encode preferential treatment, which we call *generalized GGF* (G³F). Based on G³F, we then formulate this novel fair optimization problem in deep RL. Finally, we explain the difficulties of solving the above problem and present some theoretical discussion.

4.1   G³F

G³F extends GGF by introducing an additional weight p to encode preferential treatment. The weight p is also called *importance weight*. Formally, let $p \in \Delta_D$ and $w \in \Delta_D$ be two fixed weighting vectors, the G³F is defined as follows:

$$G^3F_{p,w}(u) = \sum_i \omega_i u_i^\uparrow, \tag{8}$$

where $u \in \mathbb{R}^D$ and the weight ω_i is defined as:

$$\omega_i = w^* \left(\sum_{k=1}^{i} p_{\sigma(k)} \right) - w^* \left(\sum_{k=1}^{i-1} p_{\sigma(k)} \right), \tag{9}$$

with w^* being a monotone increasing function that linearly interpolates the points $(i/D, \sum_{k=1}^{i} w_k)$ together with the point $(0,0)$, and σ is the permutation sorting the components of vector u in increasing order, i.e., $u_{\sigma(i)} = u_i^\uparrow$ for all i.

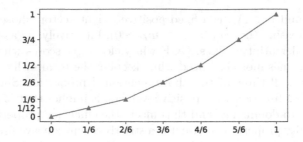

Fig. 1. Linear interpolation graph for different x values

Intuitively, such preferential treatment can be enforced via *user duplication* [6], which states that if a user is more important, s/he should be counted more times (via importance weight) than other users. Since this weight is often normalized, formally, if each user i receives some fractional entitlement p_i (i.e., importance weights), when two users are equally important, they would receive equal weights. In contrast, if a user is entitled to a preferential treatment, s/he would consequently receive a larger share of the total importance weight. The exact choice of p therefore depends on the specific problem one wants to solve.

Recall that G^3F in (8) is defined with positive decreasing weights w, it therefore satisfies efficiency, equity, and impartiality, but without assuming that all users are equal. Obviously, G^3F will reduce to GGF when p follows a uniform distribution. For a better illustration on why G^3F is a suitable choice in our setting, we consider the following example.

Example 1. Assume given an instance of $G^3F_{p,w}$ where w is chosen as $(3/6, 2/6, 1/6)$ and p is set to $(4/6, 1/6, 1/6)$. By applying linear interpolation at the key points: $w^*(0) = 0$, $w^*(1/3) = 1/6$, $w^*(2/3) = 1/2$, $w^*(1) = 1$, we can obtain the complete values (see Fig. 1). The following cases show that G^3F satisfies the three properties of a fair solution.

In the first case, We consider two vectors $u = (10, 5, 15)$, and $u' = (12, 5, 15)$. By efficiency, u' should be preferred to u, which is true by comparing the corresponding aggregation values:

$$G^3F_{p,w}(u) = 9.17, G^3F_{p,w}(u') = 10.50.$$

Let $u = (10, 5, 15)$, and $u' = (10, 12, 8)$ in the second case. By equity, u' should be preferred to u since the last two objectives are equally important and u' is more balanced over them. Indeed, we have:

$$G^3F_{p,w}(u) = 9.17, G^3F_{p,w}(u') = 9.67.$$

Therefore the equity property holds.

In the lase case, we consider $u = (10, 5, 15)$, and $u' = (10, 15, 5)$. And we can obtain:

$$G^3F_{p,w}(u) = G^3F_{p,w}(u') = 9.17.$$

Thus the two solutions are equivalent, which verifies the impartiality property.

4.2 Problem Statement

By integrating G^{3}F with MOMDPs, we can now formally formulate this fair optimization problem with preferential treatment investigated in our paper, which is the problem of determining a policy that generates a fair distribution of rewards subject to the preference weighting vector. Since we focus on deep RL, we directly write this problem with parametrized policy π_θ:

$$\underset{\pi_\theta}{\mathrm{argmax}}\, \mathrm{G^3F}_{p,w}\left(\boldsymbol{J}(\pi_\theta)\right), \tag{10}$$

where $\boldsymbol{J}(\pi_\theta)$ corresponds to the vectorial version of the standard RL objective. A solution to this problem is called G^{3}F-*fair* policy or simply *fair* policy if the context is clear. Note that both p and w are fixed and depend notably on the problem domain, its context, and what the system designer wants to achieve. These weights are therefore part of the problem description.

While the usual approaches in MOMDPs aim to find the set of Pareto optimal solutions (or an approximation), the goal of our problem is to directly learn the Pareto-optimal G^{3}F-fair policy. In addition, instead of applying G^{3}F on the immediate rewards, our formulation applies G^{3}F on the cumulative rewards over trajectories to reach more equitable reward distribution, since it allows compensation over time and expectation in this way.

4.3 Difficulties

Similarly to GGF optimization in RL [46], several challenges exist for solving Problem (10): (i) G^{3}F is a non-linear function, which makes the problem harder to solve than standard RL. However, interestingly, G^{3}F is a concave function (see Sect. 4.3), which suggests that (10) may still retain some nice properties. (ii) fair solutions may depend on initial states. (iii) stochastic policies may dominate deterministic policies when taking fairness into account.

For GGF, Siddique et al. [46] also discuss those points for the average reward criterion, and in addition, introduce an approximation bound in terms of average reward between the policy optimal for the discounted reward and that for the optimal reward. Those results can be extended to G^{3}F, but to keep the exposition simple, we do not present them in this paper.

Concavity Analysis. Although Ogryczak and Śliwiński [38] have proved the concavity of G^{3}F, here we provide another straightforward proof as an alternative.

Lemma 1. *For any $p \in \Delta_D$, for any $w \in \Delta_D$ such that its components are decreasing, function $G^3F_{p,w}$ is concave.*

Proof. Our proof relies on the following proposition proved by [25], which shows that a Choquet integral with respect to a super-modular capacity is a concave function. Thus, we only need to check that G^{3}F is such Choquet integral.

We first recall a few definitions. A *capacity* is a set function $\nu : 2^{\{1,...,D\}} \to [0,1]$ such that $\nu(\emptyset) = 0$, $\nu(\{1,...,D\}) = 1$ and $\forall A, B \subseteq \{1,...,D\}$, $A \subseteq B \implies \nu(A) \leq \nu(B)$. The *capacity* is said to be super-modular if $\nu(A \cup B) + \nu(A \cap B) \geq \nu(A) + \nu(B), \forall A, B \subseteq \{1,...,D\}$.

The *Choquet integral* [13] of a vector $\boldsymbol{v} \in \mathbb{R}^D$ with respect to a capacity ν is defined by:

$$C_\nu(\boldsymbol{v}) = \sum_{i=1}^{D} [\nu(X_{\sigma(i)}) - \nu(X_{\sigma(i+1)})]\boldsymbol{v}_{\sigma(i)}, \tag{11}$$

where $\sigma \in \mathbb{S}_D$ such that $\boldsymbol{v}_{\sigma(i-1)} \leq \boldsymbol{v}_{\sigma(i)}$ for all $i = 2,...,D$, $X_{\sigma(i)} = \{\sigma(i), \sigma(i+1),...,\sigma(D)\}$ for all $i = 1,...,D$, and $X_{\sigma(D+1)} = \emptyset$.

By identification, it is easy to see that G^{3}F is a Choquet integral with respect to the following capacity ν:

$$\nu(E) = w^*(\sum_{i \in E} \boldsymbol{p}_i), \tag{12}$$

where E is a subset of $\{1,...,D\}$.

We then show that the linear interpolation function w^* is convex when the components of $\boldsymbol{w}$ are decreasing. Since w^* is obtained by applying linear interpolation, we can easily find the slope m_i of each straight line connecting $\left(\frac{i}{D}, w^*(\frac{i}{D})\right)$ and $\left(\frac{i+1}{D}, w^*(\frac{i+1}{D})\right)$ for all $i = 0,...,D-1$,

$$m_i = \frac{\Delta y_i}{\Delta x_i} = \frac{w^*(\frac{i+1}{D}) - w^*(\frac{i}{D})}{1/D} = D\boldsymbol{w}_{D-i}. \tag{13}$$

Thus the function w^* is convex by observing that the slope m_i increases as i increases.

Next, we show that the capacity ν defined in (12) is a super-modular set function.

For any two sets $A, B \subseteq \{1,...,D\}$, and $A \cap B = C$, suppose $\sum_{i \in A} \boldsymbol{p}_i = x_1$, $\sum_{i \in B} \boldsymbol{p}_i = x_2$, $\sum_{i \in C} \boldsymbol{p}_i = x_3$, and the straight line connecting the two points $(x_3, w^*(x_3))$ and $(x_1 + x_2 - x_3, w^*(x_1 + x_2 - x_3))$ is given by $f(x) = kx + b$, then we have:

$$f(x_1 + x_2 - x_3) + f(x_3) = k(x_1 + x_2) + 2b \tag{14}$$
$$= f(x_1) + f(x_2). \tag{15}$$

Since $f(x_1 + x_2 - x_3) = w^*(x_1 + x_2 - x_3)$, $f(x_3) = w^*(x_3)$, and $f(x_1) > w^*(x_1)$, $f(x_2) > w^*(x_2)$ (w^* is convex), the following inequality therefore holds:

$$\nu(A \cup B) + \nu(A \cap B) \geq \nu(A) + \nu(B). \tag{16}$$

Therefore the capacity ν is a super-modular set function, which implies that G^{3}F is concave, by [25]'s proposition.

The concavity of G^3F implies that the optimization problem (10) has some nice properties. For instance, with a linear approximation scheme, the overall problem would be a convex optimization problem (i.e., any local optimum would be global). In deep RL, the overall problem is not convex anymore, but from the point of view of the last layer of a neural network (which is usually linear, e.g., in DQN), the optimization problem is still convex. This suggests that the overall problem is relatively well-behaved, and provides guidance on our algorithm design (see Sect. 5).

State-Dependent Optimality. As an extension of the problem investigated by [46], similarly to GGF-fair policy, an G^3F-fair policy may depend on the initial states or more generally, on the distribution of initial states. Example 2 gives an simple illustration of this property. Related to this point, because of the non-linearity of G^3F, the Bellman principle of optimality does not hold anymore and dynamic programming can not be directly applied for finding a fair optimal policy.

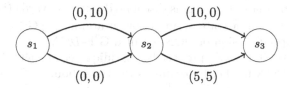

Fig. 2. Example of MDP where optimal fair solution for Problem (10) is state-dependent.

Example 2. Figure 2 depicts a deterministic bi-objective MDP with three states. Each state has two actions *(Up, Down)*, except for the terminal state s_3. Action *Up* (resp. *Down*) is represented by the arc above (resp. below). The arc weights correspond to vector rewards. Assume $p = (0.6, 0.4)$ and $w = (0.9, 0.1)$. Then if starting from s_1, the optimal policy π^* obviously selects *Up* in s_1, and *Up* again in s_2 ($G^3F_{p,w}(10, 10) = 10.0$, $G^3F_{p,w}(5, 15) = 5.8$). However, if taking s_2 as the initial state, the best action to take becomes *Down* ($(G^3F_{p,w}(10, 0) = 2.8$, $G^3F_{p,w}(5, 5) = 5.0)$), which contradicts the optimal policy π^*.

Optimality of Stochastic Policies. It is known that an optimal deterministic policy exists in the single-objective MDP setting. However, when taking fairness into account in the MOMDP setting, learning a policy only from the set of deterministic policies may not be optimal [7]. For instance, given a MOMDP with two objectives, mixing a policy optimal for the first objective with a policy optimal for the second objective can lead to a better trade-off between the two objectives (i.e., fairer solution).

5   Proposed Algorithms

In this section, we explain how to integrate G^{3}F with several existing RL algorithms (DQN, A2C, and PPO) for solving Problem (10). Notably, we introduce a novel state-augmented DQN-based method for learning stochastic policies.

5.1   Value-Based Methods

Value-based RL methods aim to estimate the optimal action-value function, namely Q_{π^*}. DQN [28] is one typical value-based deep RL method. We discuss next its extension to G^{3}F.

G^{3}F-DQN. Following [46], we modify the output of the deep Q-network to take values in $\mathbb{R}^{|\mathcal{A}| \times D}$ instead of $\mathbb{R}^{|\mathcal{A}|}$. The target Q-value is changed to:

$$\hat{Q}_\theta(s, a) = r + \gamma \hat{Q}_{\theta'}(s', a^*),$$

where $a^* = \text{argmax}_{a' \in \mathcal{A}} \, \text{G}^3\text{F}_{p,w}\left(r + \gamma \hat{Q}_{\theta'}(s', a')\right)$. The best next action is chosen such that the immediate reward plus discounted future rewards (both vectorial) is fair. For execution in a state s, an action in $\text{argmax}_{a \in \mathcal{A}} \text{G}^3\text{F}_{p,w}\left(\hat{Q}_\theta(s, a)\right)$ is chosen. This adapted version of DQN is called G^{3}F-*DQN*.

It is similar to GGF-DQN proposed by Siddique et al. [46]. Here, one may notice that G^{3}F-DQN implicitly optimizes the lower bound[1] $\mathbb{E}_{s'}\left[\text{G}^3\text{F}_{p,w}\left(r + \gamma\right.\right.$ $\left.\left.\hat{Q}_{\theta'}(s', a_s^*))\right]\right]$ instead of $\text{G}^3\text{F}_{p,w}\left(\mathbb{E}_{s'}\left[r + \gamma\hat{Q}_{\theta'}(s', a_s^*)]\right)\right.$, which would be a better approximation of the objective function of (10). For this reason, one may not expect a very good performance from G^{3}F-DQN. Next, we propose two other novel extensions of DQN that can achieve better performance.

G^{3}F-CDQN. Recall that an optimal fair policy may depend on initial states (Sect. 4.3), and that in G^{3}F-DQN, the learned policy is both deterministic and Markov, which is not sufficient to achieve fairness in an effective way. While still aiming for a deterministic policy here, a natural approach to address the other two points is state augmentation. Indeed, if the agent can base its decisions on both past accumulated reward and usual state information, the agent may be able to achieve a higher level of fairness. Intuitively, such additional information enables the agent to base its decisions on past accumulated reward, which can help correct past inequities.

Consequently, we first augment an original state s_t as follows:

$$\bar{s}_t = \left(s_t, \frac{1}{\lambda}r_{1:t}\right)$$

---

[1] $\mathbb{E}_{s'}\left[\text{G}^3\text{F}_{p,w}\left(r + \gamma\hat{Q}_{\theta'}(s', a_s^*))\right)\right]$ is a lower bound of $\text{G}^3\text{F}_{p,w}\left(\mathbb{E}_{s'}\left[r + \gamma\,\hat{Q}_{\theta'}(s', a_s^*)\right]\right)$ by Jensen inequality since $\text{G}^3\text{F}_{p,w}$ is concave. Notation a_s^* is to emphasize its dependence on s.

where $\lambda = \sum_{\tau=1}^{t-1} \gamma^{\tau-1}$ acts as a scaling factor, $\boldsymbol{r}_{1:t} = \sum_{\tau=1}^{t-1} \gamma^{\tau-1} \boldsymbol{r}_\tau$ denotes the discounted cumulative reward received so far, which is reset to zero at the beginning of an episode. Then we modify the target Q-value as follows:

$$\hat{\boldsymbol{Q}}_\theta(\bar{s}_t, a) = \boldsymbol{r}_t + \gamma \hat{\boldsymbol{Q}}_{\theta'}(\bar{s}_{t+1}, a^*),$$

where $a^* = \text{argmax}_{a' \in \mathcal{A}} G^3 F_{p,w} (\hat{\boldsymbol{Q}}_{\theta'}(\bar{s}_{t+1}, a'))$. Here the immediate reward r_t is removed from the G^3F computation since this signal is already included in the augmented state as part of the cumulative reward. For execution in a state s, an action in $\text{argmax}_{a \in \mathcal{A}} G^3 F_{p,w} (\hat{\boldsymbol{Q}}_\theta(\bar{s}, a))$ is chosen. This algorithm is called G^3F-*CDQN*.

G^3F-CSDQN Since stochastic policies may dominate deterministic ones (Sect. 4.3), we may improve the performance of G^3F-CDQN by learning a stochastic policy. We describe how to achieve this next.

First, we describe how G^3F-DQN can be modified to learn stochastic policies. The target Q-value is changed to:

$$\hat{\boldsymbol{Q}}_\theta(s, a) = \boldsymbol{r} + \gamma \hat{\boldsymbol{Q}}^*_{\theta'}(s', \cdot),$$

where $\hat{\boldsymbol{Q}}^*_{\theta'}(s', \cdot) = \sum_{a' \in \mathcal{A}} \pi^*(a'|s') \hat{\boldsymbol{Q}}_{\theta'}(s', a')$ denotes an estimated Q-value achieved at a next state by a policy π^*, which is defined as:

$$\pi^*(\cdot|s') = \text{argmax}_\pi G^3 F_{p,w} \left(\boldsymbol{r} + \gamma \sum_{a' \in \mathcal{A}} \pi(a'|s') \hat{\boldsymbol{Q}}_{\theta'}(s', a') \right) \tag{17}$$

This reformulation assumes that in the next state, the best stochastic policy is chosen (in contrast to the deterministic greedy policy in DQN or G^3F-DQN). For execution in a state s, an action is sampled from $\pi^*(\cdot|s)$ obtained from $\text{argmax}_\pi G^3 F_{p,w} (\sum_{a' \in \mathcal{A}} \pi(a'|s') \hat{\boldsymbol{Q}}_\theta(s', a'))$.

Problem (17) is a non-linear convex optimization problem that can be solved via linear programming [37]:

$$max \sum_{k=1}^{D} \frac{k}{n} w'_k x_k - \sum_{k=1}^{D} \sum_{i=1}^{D} w'_k p_i d_{ik} \tag{18}$$

$$s.t. \; x_k - d_{ik} \leq r_i + \gamma y_i, \; \forall i, k = 1, \ldots, D$$

$$\boldsymbol{y} = \sum_{a' \in \mathcal{A}} \pi(a'|s') \hat{\boldsymbol{Q}}_{\theta'}(s', a')$$

$$0 \leq \pi(a'|s') \leq 1, \; \sum_{a' \in \mathcal{A}} \pi(a'|s') = 1$$

$$d_{ik} \geq 0, \; \forall i, k = 1, \ldots, D$$

where $w'_k = D(w_k - w_{k+1})$ for $k = 1, \ldots, D-1$, $w'_D = D w_D$, x_k's and d_{ik}'s are additional variables introduced to linearize the original non-linear optimization problem. Finally, by introducing state augmentation like in G^3F-CDQN, we can formulate a novel algorithm called G³F-$CSDQN$, which can learn fair stochastic policies for augmented states. Note that although one may expect a better performance from this new algorithm, G³F-CDQN may still be useful in domains where deterministic policies are favored (e.g., robotics).

5.2   Policy Gradient Methods

Although usually less sample-efficient than DQN-based algorithms, policy gradient methods constitute another natural choice for solving Problem (10). Following the work by Siddique et al. [46], we show how to extend two actor-critic (AC) methods: A2C [27] and PPO [43] to solve our problem. We call our new algorithms: G³F-A2C and G³F-PPO respectively. A nice feature of those methods is that they can directly learn a stochastic policy. Note that other policy gradient methods could be extended in a similar fashion.

G³F-**A2C.** To reduce the variance of the estimation of the policy gradient (5), A2C uses a control variate method where a state-dependent baseline is subtracted from $\boldsymbol{Q}_{\pi_\theta}$. Using $v(s)$ as a baseline yields the advantage function, which is estimated in A2C by $\boldsymbol{A}_{\mathrm{A2C}}(s_t, a_t) = \sum_{t=1} \gamma^{t-1} R_t - v(s_t)$ where R_t is the immediate reward obtained at time step t. In A2C, the actor update derives from the policy gradient obtained from:

$$\boldsymbol{J}_{\mathrm{A2C}}(\boldsymbol{\theta}) = \mathbb{E}_{s \sim d_\pi, a \sim \pi_\theta(\cdot|s)}[\boldsymbol{A}_{\mathrm{A2C}}(s, a)].$$

For G³F-A2C, the policy gradient is formulated as follows:

$$\nabla_\theta \mathrm{G^3F}_{\boldsymbol{p},\boldsymbol{w}}(\boldsymbol{J}_{\mathrm{A2C}}(\boldsymbol{\theta})) \tag{19}$$

$$= \nabla_{\boldsymbol{J}_{\mathrm{A2C}}(\theta)} \mathrm{G^3F}_{\boldsymbol{p},\boldsymbol{w}}(\boldsymbol{J}_{\mathrm{A2C}}(\boldsymbol{\theta})) \cdot \nabla_\theta \boldsymbol{J}_{\mathrm{A2C}}(\boldsymbol{\theta}) \tag{20}$$

$$= \boldsymbol{\omega}_\sigma^\mathsf{T} \cdot \nabla_\theta \boldsymbol{J}_{\mathrm{A2C}}(\boldsymbol{\theta}), \tag{21}$$

where $\nabla_{\boldsymbol{J}_{\mathrm{A2C}}(\theta)} \mathrm{G^3F}_{\boldsymbol{p},\boldsymbol{w}}(\boldsymbol{J}_{\mathrm{A2C}}(\boldsymbol{\theta})) \in \mathbb{R}^D$ is the gradient of function $\mathrm{G^3F}_{\boldsymbol{p},\boldsymbol{w}}$ with respect to its components and $\nabla_\theta \boldsymbol{J}_{\mathrm{A2C}}(\boldsymbol{\theta}) \in \mathbb{R}^{D \times N}$ (N being the number of policy parameters) represents the classic policy gradient extended to the vector case.

G³F-**PPO.** Following the design of PPO [43], the advantage is estimated with λ-returns. Formally, the estimated advantage function $\boldsymbol{A}_{\mathrm{PPO}}(s, a)$ can be written as $\boldsymbol{A}_{\mathrm{PPO}}(s_t, a_t) = \sum_t (\gamma\lambda)^{t-1} \delta_t$ where $\delta_t = R_t + \gamma v(s_{t+1}) - v(s_t)$. A similar clipped surrogate objective function can be formulated to guide policy training. Denoted $\boldsymbol{J}_{\mathrm{PPO}}(\boldsymbol{\theta})$, it is defined so as to limit policy changes after an update:

$$\mathbb{E}_{s \sim d_\pi, a \sim \pi_\theta(\cdot|s)}[\min(\rho_\theta \boldsymbol{A}_{\mathrm{PPO}}(s, a), \bar{\rho}_\theta \boldsymbol{A}_{\mathrm{PPO}}(s, a))], \tag{22}$$

where $\rho_\theta = \frac{\pi_\theta(a|s)}{\pi_b(a|s)}$ is an importance sampling weight, π_b is the behavior policy generating the training data, $\bar{\rho}_\theta = \mathrm{clip}(\rho_\theta, 1 - \epsilon, 1 + \epsilon)$ is a clipped weight, and ϵ is a hyperparameter to control how much the current policy can change. The policy gradient for G³F-PPO can then be obtained by replacing $\boldsymbol{J}_{\mathrm{A2C}}$ in (19) with $\boldsymbol{J}_{\mathrm{PPO}}$.

6   Experimental Results

To validate the effectiveness of our proposed methods, we conducted comprehensive experiments to evaluate our algorithms (with relevant baselines) across three distinct domains. In those domains, our primary objective is to assess the performance of our approach, both in terms of achieving fairness and maintaining desirable learning outcomes. To evaluate our approach, we set the weights $w_i = \frac{1}{2^i}, i = 0, ..., D - 1$. It is worth noting that, we carried out experiments with different p configurations to evaluate the influence on the G^3F algorithms. Importantly, we assume that the weights p hold to the condition $\sum_i p_i = 1$. It is imperative to emphasize that weights p and w are problem-dependent and should be set by the system designer to achieve the level of fairness s/he desires. Moreover, to ensure that our results are reproducible, we average the results over 10+ runs, each initialized with different random seeds across all domains. For hyperparameter optimization, we use an open-source library called Lightweight HyperParameter Optimizer (LHPO) [57]. LHPO is designed for parallelized experimentation, enabling us to present results with the most optimal set of hyperparameters. In the following sections, we present a detailed description of each domain and show the results of our experiments.

6.1   Species Conservation

Our first experimental domain is a species conservation (SC) problem, which constitutes a critical domain in the field of ecology, particularly when dealing with the task of preserving multiple interacting endangered species. In this context, we address the challenge of introducing fairness considerations into the conservation efforts of two endangered species, sea otters and abalones. Naturally, sea otters live on a diet consisting primarily of abalones, which creates an intricate ecological balance—maintaining a sufficient abalone population for sea otters' survival. Compounding this natural challenge, human activities, including predation and oil contamination, pose severe threats to the sea otter population. On the other hand, abalones are subject to both natural mortality and predation by sea otters, demanding strict managerial actions to ensure the coexistence and preservation of both species. To simulate such a complex conservation problem, we adopt the framework proposed by [8]. Within this problem, the state is comprised of the population level of the two species. The action space consists of five actions including *do nothing, introduce sea otters, enforce antipoaching, control sea otters*, and *one-half antipoaching and one-half control sea otters*. Each of these managerial actions exerts its unique influence. For instance, introducing sea otters can reduce the abalone population, as sea otters will consume more of them. However, excessive introductions may drive abalones to extinction, disrupting the ecological balance. Similarly, attempts to control sea otters can lead to unintended consequences, causing the abalone population to rise while leading the sea otter population to extinction. The transition function in this framework incorporates population growth dynamics while also accounting for the impact of antipoaching and oil spills. The reward function in this framework is composed of each species' density. Since we express the fairness over the two species $(D = 2)$ this can be understood as both species remaining alive and having a balanced population. Because densities may not be comparable directly, using equal weights for p may not be suitable. In that case, G^3F may be beneficial.

In this domain, our goal is to rigorously assess the effectiveness of our proposed methods in optimizing the G^3F function. Our central question revolves around determining whether the optimization of these G^3F scores genuinely ensures balanced solutions across both species. Additionally, we aim to investigate how various configurations of p influence the performance of G^3F algorithms, particularly in cases when preferential treatment is important for certain users while concurrently maintaining a fair solution for all other users. The experimental results and their analysis are conducted with the goal of addressing the following questions:

1. Does G^3F algorithms yield higher G^3F score than GGF or standard algorithms?
2. What is the effect of training a policy with different weights for p?
3. Does considering past discounted reward or learning a stochastic policy help in DQN-based algorithms?

Does G^3F Algorithms Yield Higher G^3F Score than GGF or Standard Algorithms? This first question is a sanity check to verify that our new algorithms do optimize G^3F. We compare the G^3F scores of DQN, A2C, and PPO with their GGF and G^3F counterparts in the SC domain. The G^3F scores are obtained by applying $G^3F_{p,w}$ on the empirical average vector returns of trajectories sampled with the learned policies during the test phase. Figure 3 depicts the distribution of this score for the policies learned by DQN, A2C, PPO, and their GGF and G^3F extensions. The weight p is set to $(0.9, 0.1)$ for G^3F methods, where the first component corresponds to sea otters and the second one corresponds to the abalones. Our results reveal that the GGF algorithms can find a fairer solution than their original versions, and thus have a higher G^3F score. However, the G^3F algorithms show an even higher score than both their GGF and original counterparts, indicating that fairness with priority set by p was better achieved, as can be seen in Fig. 4. For instance, the G^3F-A2C variant nearly balances the population densities, primarily assigning the higher priority given to sea otters. Recall that, in nature, the density of abalones is inherently much larger [46].

As the G^3F score does not directly show the vector compositions, plots of non-aggregated accumulated densities estimated during the testing phase are also presented (Fig. 4), which is simple to do for the SC domain because it is bi-objective. Compared to the standard or GGF counterparts, optimizing G^3F with a higher priority given to sea otters achieves more balanced individual densities. This suggests that a non-uniform p may help correct advantages conferred to some users by the environment.

What is the Effect of Training a Policy with Different Weights for p? To answer this question, we evaluate the performances of the G^3F algorithms

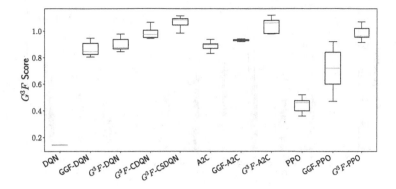

Fig. 3. G^3F scores of DQN, A2C, PPO, and their GGF or G^3F counterparts in SC during the testing phase. Weight $p = (0.9, 0.1)$ for the G^3F algorithms.

with different weights for p in the domain. Figure 5 shows the performance of G^3F-CDQN, G^3F-CSDQN, G^3F-A2C, and G^3F-PPO during the testing phase in terms of *Coefficient of Variation* (CV), minimum and maximum density. Recall that CV is defined as the ratio of the standard deviation to the mean. It can be interpreted as a simple measure of inequality, with lower CV values implying more balanced solutions. Given our prior findings, where GGF and G^3F algorithms exhibited more balanced reward distributions, characterized by lower CV values and higher minimum density, this experiment seeks to shed light on the consequences of changing the preference weight associated with the first objective. For experiments in the SC domain, we increase the preference weight of the first objective p_0 from 0.1 to 0.9 (i.e., p_1 decreases from 0.9 to 0.1, correspondingly). By assigning the weight to the first objective, we effectively prioritize sea otters. As a result, this leads to a higher sea otter population density, which in turn results in lower CV values, higher minimum density, and lower maximum density for all G^3F algorithms. This result provides valuable insights into the ability of our proposed G^3F algorithms to assign preferential treatment in cases where one objective naturally assumes a significantly higher or lower standing than the other, ultimately leading to a more balanced solution.

Does Considering Past Discounted Reward or Learning a Stochastic Policy Help in DQN-Based Algorithms? To address the question of whether incorporating past discounted rewards or learning a stochastic policy is advantageous to DQN-based algorithms, we conduct a comprehensive comparison of all our DQN variants within the SC domain. Figs. 6, 7 and 3 show the performances of those algorithms in the SC domain. Notably, while Fig. 6 illustrates the training curves within 60k interactions, the AC methods are indeed trained with 600k interactions for convergence before testing. These figures collectively show a consistent trend that moving from DQN, G^3F-DQN, G^3F-CDQN, to G^3F-CSDQN nearly always yields an increase in terms of average density (more efficient), a decrease in terms of CV (more equitable), an increase in terms of min density (more equitable), and an increase in terms of G^3F (fairer). This latter

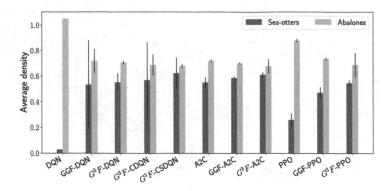

Fig. 4. Population densities of DQN, A2C, PPO, and their GGF or G³F counterparts in SC during the testing phase. Weight $p = (0.9, 0.1)$ for the G³F algorithms.

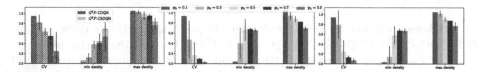

Fig. 5. Effects of using different weights for p in G³F algorithms. CV, minimum and maximum densities of G³F-CDQN and G³F-CSDQN (left), G³F-A2C (middle), and G³F-PPO (right) during testing in SC.

point experimentally confirms the theoretical discussion about the optimality of stochastic policies in Sect. 4.3.

Interestingly, G³F-CDQN and G³F-CSDQN outperform DQN in terms of average density, which is exactly what is optimized by DQN. This is explained by the fact that this domain is actually partially observable. In addition, Fig. 6 also includes the training curves of the AC methods (A2C and PPO) for comparison. It can be observed that the DQN-based variants learn much faster than the AC methods in terms of the number of interactions. Therefore, the DQN-based variants would become more preferable choices when the sample efficiency is important.

While the above results are obtained with non-uniform p, we extend our investigation to the GGF setting, characterized by uniform p. Remarkably, our G³F-CDQN (i.e., GGF-CDQN) outperforms GGF-DQN in finding better solutions, suggesting the adaptability of our novel DQN-based methods in addressing standard fair optimization problems without any preferential treatment.

To provide additional insights into the relationship between lower CV values and more balanced reward distributions, we present individual density plots for varying p values in Fig. 8, depicting the population densities of both species for G³F-CDQN and G³F-CSDQN under varying p values. This plot illustrates how the population densities of both species evolve under different algorithms and with varying p configurations. Again our results demonstrate that increasing

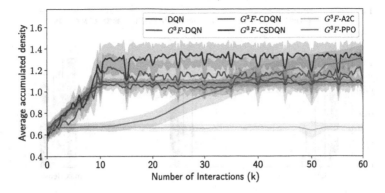

Fig. 6. Average accumulated densities of DQN-based algorithms, G³F-A2C, and G³F-PPO in SC during training. The weight p is set to $(0.9, 0.1)$ for G³F algorithms.

the weight p_0 leads to a more equitable distribution of population densities for both species. Moreover, G³F-CSDQN tends to perform better than G³F-CDQN, consistent with our previous observations.

6.2 Traffic Light Control

Our second experimental domain is a traffic light (TL) control, a real-world problem that presents a set of challenges due to its inherent complexity and the multifaceted nature of its objectives. In this problem, an autonomous agent controls the traffic lights at a single intersection, with the goal of optimizing the flow of traffic. To simulate this real-world traffic behavior, we use the Simulation of Urban Mobility (SUMO)[2]. Within this domain, our focus is on the 8-lane standard single intersection, with two lanes designated for each side of the road. Traditionally, the primary goal in this domain is to minimize the total waiting time of all cars stopped at the intersection. However, we consider fairness over each direction at the intersection (i.e., $D=4$). Here, the state in this domain is composed of the waiting times and densities of cars waiting at the intersection. An action amounts to selecting the next traffic-light phase. We assumed four phases: NSL, NSSR, EWL, and EWSR, with NSR representing the (north-south left) phase when the green light is assigned to the left lanes of roads approaching from the north and south, NSSR representing the (north-south straight and right) phase when the green light is assigned to the straight and right lanes of roads approaching from the north and south, and so on. In this problem, while giving equal treatment to all objectives is needed, we additionally also consider some behaviors where some lanes will be given preferential treatment (e.g., due to morning rush, traffic flows are unbalanced) and that the waiting times for cars in these lanes will be optimized with higher priorities, while other lanes with equal preferences will be treated fairly. Through experiments in this domain, we seek answers to the following questions:

---

[2] https://github.com/eclipse/sumo.

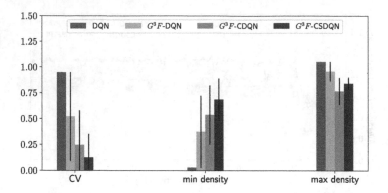

Fig. 7. CV, minimum and maximum densities of DQN-based algorithms in SC during testing. The weight p is set to $(0.9, 0.1)$ for G^3F algorithms.

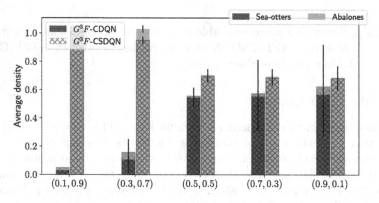

Fig. 8. Population densities of G^3F-CSDQN and G^3F-CDQN algorithms in SC during testing phase with varying p.

1. What is the impact of learning a stochastic policy with past accrued reward in DQN-based methods for TL control?
2. How much control over solutions does p provide in the TL domain?

What is the Impact of Learning a Stochastic Policy with Past Accrued Reward in DQN-Based Methods for TL Control? To address the question, we investigate the effects of learning a stochastic policy and accrued rewards on the performance of DQN-based methods within the TL domain by presenting results with and without the inclusion of stochastic policy and accrued rewards to show the significance of each component. First, we show the training performance of DQN, G^3F-DQN, G^3F-CDQN, G^3F-CSDQN, G^3F-A2C, and G^3F-PPO in Fig. 9. To show the sample efficiency of our methods, we limit the displayed training results to the first 60,000 interactions. Note that the x-axis corresponds to the number of interactions, which may not correspond to the timesteps in an environment (e.g., A2C simultaneously use several environments to generate training data). These results clearly show that DQN-based methods learn much faster than the AC methods in terms of the number of interactions. Therefore, the DQN-based variants would become more preferable choices when the sample efficiency is important. Once again, G^{3}F-CSDQN outperforms or achieves similar results with their DQN variants in terms of average accumulated waiting time, showing that even when attempting to strike a balance between fairness and efficiency, G^{3}F-CSDQN consistently maintains a high level of operational efficiency.

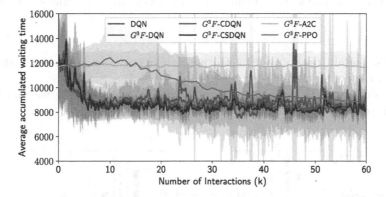

Fig. 9. Learning performance of DQN, G^{3}F-DQN, G^{3}F-CSDQN, G^{3}F-CDQN,G^{3}F-A2C, and G^{3}F-PPO in TL during training phase.

To further confirm the effectiveness of our proposed methods, we also assess our algorithms based on several essential metrics, including the CV, minimum waiting time, and maximum waiting time, as displayed in Fig. 10. Repeatedly, our results indicate that the transition from DQN, G^{3}F-DQN, G^{3}F-CDQN, to G^{3}F-CSDQN nearly always yields a consistent trend: a decrease in CV, an increase in minimum waiting time, a decrease in maximum waiting time, and an increase in the G^{3}F score. Notably, our G^{3}F-CDQN manages to outperform other DQN-based methods, showing the efficacy of our methods in learning efficient fair solutions.

How Much Control Over Solutions Does p Provide in the TL Domain? In the TL domain, we vary weight p_0 (assigned to North), while the remaining weight is assigned uniformly over the remaining three components (directions) of p. As shown in Fig. 11, waiting times of cars coming from lanes with higher weights are shorter than those coming from lanes with lower weights. It can be observed that the waiting times of cars coming from the north and south are close, despite the fact that they are assigned different weights. This is due to the fact that the agent's action can affect two lanes at the same time in this case. For example, an action *NSL* corresponds to the phase when the left lanes of north and south are given a green light and cars can only turn left during this

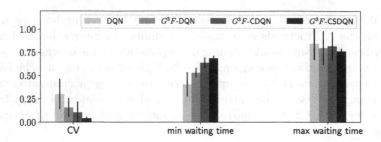

Fig. 10. CV, minimum and maximum densities of DQN-based algorithms in TL during testing.

phase. As a result, optimizing the waiting time in one lane will have an effect on the opposite lane as well.

The above results show that by appropriately adjusting the weights p, we can achieve the desired control over multiple objectives.

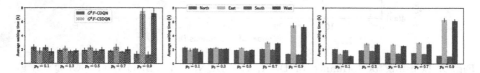

Fig. 11. Effects of using different weights for p in G^3F algorithms. Individual waiting times of G^3F-CDQN and G^3F-CSDQN (left), G^3F-A2C (middle), and G^3F-PPO (right) during testing in TL.

6.3 Data Center

Our third and last domain is the data center (DC) traffic congestion control problem [42], which involves connecting a large number of computers according to some network topology. In particular, we consider a network with a fat-tree topology, which connects 16 computers via 20 switches. To simulate the dynamics of this fat-tree network topology, we used Mininet[3], with UDP serving as the underlying transport protocol. For generating traffic and gathering information, we employed Goben[4]. Traditionally, the goal in this problem entails the learning of a controller capable of maximizing the cumulative host bandwidth, while accounting for queue lengths to avoid the switch bufferbloats. However, we instead aim to optimize the bandwidth for each host (e.g., $D=16$), an effective way to impose fairness criteria within this domain. Here, the state is composed of each computer network information, encompassing statistics from transport and lower layers. A continuous action corresponds to the allocation of bandwidth

---

[3] https://github.com/mininet/mininet.
[4] https://github.com/udhos/goben.

for each host. The vector reward is calculated by penalizing the bandwidths per host by the sum of queue lengths. In this domain, our goal extends beyond just treating all users equally. By intentionally assigning non-uniform preferential treatment to certain objectives, we intend to examine the effectiveness of G^3F-based algorithms and their ability to handle diverse scenarios.

When More Weight is Given to One Objective and Equal Weights are Assigned to the Other Objectives, Does the "Equal Treatment of Equals" Principle Still Hold? The previous discussion in the TL domain suggests that this may not always be the case, due to the inherent structure of the control problem. It is however interesting to answer this question when less or no dependence between objectives is expected. We therefore turn to the DC domain, where there are 16 objectives in total. In this domain, the first objective is given a weight of $\frac{1}{4}$, and the other objectives are assigned equal weights, i.e., $\frac{1}{20}$. Figure 12 illustrates the performances of standard deep RL algorithms and their GGF/G^3F counterparts in terms of CV (w.r.t the objectives with identical weights, i.e., the first objective is excluded in this statistic), minimum, and maximum bandwidths.

As expected, the GGF algorithms have a lower CV than standard RL algorithms, which indicates that they can find fairer policies than their original versions. Compared to standard or GGF versions of A2C and PPO, the G^3F counterparts have lower minimum and maximum bandwidths since more weight is given to the first objective. However, we notice that the objectives with identical weights are treated fairly, as indicated by lower CVs than standard RL algorithms, which validates that the "equal treatment of equals" principle still holds in this case.

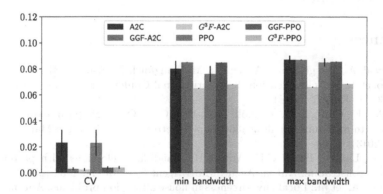

Fig. 12. CV (w.r.t the objectives with equal importance weights), minimum and maximum bandwidths of A2C, PPO and their GGF, G^3F counterparts during testing in DC.

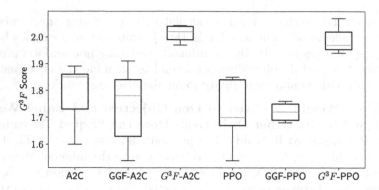

Fig. 13. G^3F scores of A2C, PPO and their GGF, G^3F counterparts during testing in DC.

7 Conclusion

We investigated the fair optimization problem with preferential treatment in RL. We presented several extensions of deep RL algorithms to tackle it, and notably proposed a novel state-augmented DQN-based method, which can be adapted to learn either deterministic or stochastic policies. Extensive experimental results on several domains were provided for validation. As future work, we plan to investigate the multi-agent extension of our new problem.

Acknowledgements. This work is supported in part by the program of the Shanghai NSF (No. 19ZR1426700).

References

1. Agarwal, A., Beygelzimer, A., Dudík, M., Langford, J., Wallach, H.: A reductions approach to fair classification. In: International Conference on Machine Learning, pp. 60–69. PMLR (2018)
2. Amaldi, E., Coniglio, S., Gianoli, L.G., Ileri, C.U.: On single-path network routing subject to max-min fair flow allocation. Electron. Notes Discrete Math. **41**, 543–550 (2013)
3. Bei, X., Liu, S., Poon, C.K., Wang, H.: Candidate selections with proportional fairness constraints. Auton. Agent. Multi-Agent Syst. **36**(1), 1–32 (2022)
4. Beynier, A., et al.: Local envy-freeness in house allocation problems. Auton. Agent. Multi-Agent Syst. **33**(5), 591–627 (2019)
5. Brams, S.J., Taylor, A.D.: Fair Division: From Cake-Cutting to Dispute Resolution. Cambridge University Press, Cambridge, March 1996
6. Brams, S.J., Taylor, A.D.: Fair Division: From Cake-Cutting to Dispute Resolution. Cambridge University Press, Cambridge (1996)
7. Busa-Fekete, R., Szörényi, B., Weng, P., Mannor, S.: Multi-objective bandits: Optimizing the generalized GINI index. In: International Conference on Machine Learning, pp. 625–634. PMLR (2017)

8. Chadès, I., Curtis, J.M., Martin, T.G.: Setting realistic recovery targets for two interacting endangered species, sea otter and northern abalone. Conserv. Biol. **26**(6), 1016–1025 (2012)
9. Chakraborty, M., Igarashi, A., Suksompong, W., Zick, Y.: Weighted envy-freeness in indivisible item allocation. ACM Trans. Econ. Comput. (TEAC) **9**(3), 1–39 (2021)
10. Chen, J., Wang, Y., Lan, T.: Bringing fairness to actor-critic reinforcement learning for network utility optimization. In: INFOCOM (2021)
11. Chevaleyre, Y., Dunne, P.E., Lemaître, M., Maudet, N., Padget, J., Phelps, S., Rodríguez-aguilar, J.A.: Issues in multiagent resource allocation. Computer **30**, 3–31 (2006)
12. Chierichetti, F., Kumar, R., Lattanzi, S., Vassilvitskii, S.: Fair clustering through fairlets. In: Advances in Neural Information Processing Systems, vol. 30. Curran Associates, Inc. (2017)
13. Choquet, G.: Theory of capacities. In: Annales de l'institut Fourier, vol. 5, pp. 131–295 (1954)
14. Chow, Y., Ghavamzadeh, M.: Algorithms for CVaR optimization in MDPs (2014)
15. Cousins, C.: An axiomatic theory of provably-fair welfare-centric machine learning. Adv. Neural. Inf. Process. Syst. **34**, 16610–16621 (2021)
16. de la Cruz, G., Peng, B., Lasecki, W.S., Taylor, M.E.: Generating real-time crowd advice to improve reinforcement learning agents. In: AAAI Workshop Learning for General Competency in Video Games, pp. 17–20 (2015)
17. Do, V., Usunier, N.: Optimizing generalized GINI indices for fairness in rankings. In: Proceedings of the 45th International ACM SIGIR Conference on Research and Development in Information Retrieval, pp. 737–747 (2022)
18. Dwork, C., Hardt, M., Pitassi, T., Reingold, O., Zemel, R.: Fairness through awareness. In: Proceedings of the 3rd Innovations in Theoretical Computer Science Conference, pp. 214–226, January 2012
19. Ghalme, G., Nair, V., Patil, V., Zhou, Y.: Long-term resource allocation fairness in average markov decision process (AMDP) environment. In: Proceedings of the 21st International Conference on Autonomous Agents and Multiagent Systems, pp. 525–533 (2022)
20. Heidari, H., Ferrari, C., Gummadi, K., Krause, A.: Fairness behind a veil of ignorance: a welfare analysis for automated decision making. In: Advances in Neural Information Processing Systems, vol. 31 (2018)
21. Jabbari, S., Joseph, M., Kearns, M., Morgenstern, J., Roth, A.: Fairness in reinforcement learning. In: International Conference on Machine Learning, pp. 1617–1626. PMLR (2017)
22. Jiang, J., Lu, Z.: Learning fairness in multi-agent systems. In: Advances in Neural Information Processing Systems, vol. 32 (2019)
23. Konidaris, G., Kuindersma, S., Barto, A., Grupen, R.: Constructing skill trees for reinforcement learning agents from demonstration trajectories. In: NIPS (2010)
24. Liu, Y., Koenig, S.: Risk-sensitive planning with one-switch utility functions: Value iteration. In: AAAI, pp. 993–999. AAAI (2005)
25. Lovász, L.: Submodular functions and convexity. In: Bachem, A., Korte, B., Grötschel, M. (eds.) Mathematical Programming the State of the Art, pp. 235–257. Springer, Berlin (1983). https://doi.org/10.1007/978-3-642-68874-4_10
26. Mandal, D., Gan, J.: Socially fair reinforcement learning. arXiv preprint arXiv:2208.12584 (2022)
27. Mnih, V., et al.: Asynchronous methods for deep reinforcement learning. In: ICML (2016)

28. Mnih, V., et al.: Human-level control through deep reinforcement learning. Nature **518**, 529–533 (2015)
29. Moulin, H.: Fair Division and Collective Welfare. MIT Press, Cambridge (2004)
30. Nabi, R., Malinsky, D., Shpitser, I.: Learning optimal fair policies. In: ICML (2019)
31. Nardi, L., Stachniss, C.: Uncertainty-aware path planning for navigation on road networks using augmented MDPs. In: ICRA (2019)
32. Nath, S., Baranwal, M., Khadilkar, H.: Revisiting state augmentation methods for reinforcement learning with stochastic delays. In: CIKM (2021)
33. Neidhardt, A., Luss, H., Krishnan, K.R.: Data fusion and optimal placement of fixed and mobile sensors. In: 2008 IEEE Sensors Applications Symposium, February 2008
34. Nguyen, V.H., Weng, P.: An efficient primal-dual algorithm for fair combinatorial optimization problems. In: COCOA (2017)
35. Ogryczak, W., Luss, H., Pióro, M., Nace, D., Tomaszewski, A.: Fair optimization and networks: a survey. J. Appl. Math. **2014** (2014)
36. Ogryczak, W., Perny, P., Weng, P.: A compromise programming approach to multiobjective markov decision processes. Int. J. Inf. Technol. Decis. Making **12**(05), 1021–1053 (2013)
37. Ogryczak, W., Śliwiński, T.: On optimization of the importance weighted OWA aggregation of multiple criteria. In: Gervasi, O., Gavrilova, M.L. (eds.) ICCSA 2007. LNCS, vol. 4705, pp. 804–817. Springer, Heidelberg (2007). https://doi.org/10.1007/978-3-540-74472-6_66
38. Ogryczak, W., Śliwiński, T.: On solving optimization problems with ordered average criteria and constraints. In: Fuzzy Optimization: Recent Advances and Applications, pp. 209–230 (2010)
39. Perny, P., Weng, P., Goldsmith, J., Hanna, J.: Approximation of lorenz-optimal solutions in multiobjective Markov decision processes. In: AAAI - Late Breaking Paper (2013)
40. Puterman, M.: Markov decision processes: discrete stochastic dynamic programming. Wiley (1994)
41. Rawls, J.: The Theory of Justice. Havard University Press, Cambridge (1971)
42. Ruffy, F., Przystupa, M., Beschastnikh, I.: Iroko: a framework to prototype reinforcement learning for data center traffic control. In: Workshop on ML for Systems at NeurIPS (2019). arxiv.org/abs/1812.09975
43. Schulman, J., Wolski, F., Dhariwal, P., Radford, A., Klimov, O.: Proximal policy optimization algorithms. CoRR abs/1707.06347 (2017). arxiv.org/abs/1707.06347
44. Sharifi-Malvajerdi, S., Kearns, M., Roth, A.: Average individual fairness: algorithms, generalization and experiments. In: Advances in Neural Information Processing Systems (2019)
45. Shi, H., Prasad, R.V., Onur, E., Niemegeers, I.G.M.M.: Fairness in wireless networks:issues, measures and challenges. IEEE Commun. Surv. Tutorials **16**(1), 5–24 (2014)
46. Siddique, U., Weng, P., Zimmer, M.: Learning fair policies in multi-objective (deep) reinforcement learning with average and discounted rewards. In: ICML (2020)
47. Singh, A., Joachims, T.: Policy learning for fairness in ranking. In: Advances in Neural Information Processing Systems (2019)
48. Sootla, A., et al.: Sauté rl: almost surely safe reinforcement learning using state augmentation. In: ICML (2022)

49. Speicher, T., et al.: A unified approach to quantifying algorithmic unfairness: measuring individual & group unfairness via inequality indices. In: Proceedings of the 24th ACM SIGKDD International Conference on Knowledge Discovery & Data Mining, pp. 2239–2248 (2018)
50. Sun, A., Chen, B., Doan, X.V.: Connections between fairness criteria and efficiency for allocating indivisible chores. arXiv preprint arXiv:2101.07435 (2021)
51. Sutton, R.S., McAllester, D., Singh, S., Mansour, Y.: Policy gradient methods for reinforcement learning with function approximation. In: NIPS (2000)
52. Van Moffaert, K., Nowé, A.: Multi-objective reinforcement learning using sets of pareto dominating policies. J. Mach. Learn. Res. **15**(1), 3483–3512 (2014)
53. Wen, M., Bastani, O., Topcu, U.: Algorithms for fairness in sequential decision making. In: ICML (2021)
54. Weng, P.: Fairness in reinforcement learning. arXiv preprint arXiv:1907.10323 (2019)
55. Zafar, M.B., Valera, I., Rodriguez, M.G., Gummadi, K.P., Weller, A.: From Parity to Preference-based Notions of Fairness in Classification. In: Advances in Neural Information Processing Systems (2017)
56. Zhang, X., Liu, M.: Fairness in learning-based sequential decision algorithms: a survey. In: Vamvoudakis, K.G., Wan, Y., Lewis, F.L., Cansever, D. (eds.) Handbook of Reinforcement Learning and Control. SSDC, vol. 325, pp. 525–555. Springer, Cham (2021). https://doi.org/10.1007/978-3-030-60990-0_18
57. Zimmer, M.: Apprentissage par renforcement developpemental. Ph.D. thesis, University of Lorraine, January 2018
58. Zimmer, M., Glanois, C., Siddique, U., Weng, P.: Learning fair policies in decentralized cooperative multi-agent reinforcement learning. In: International Conference on Machine Learning, pp. 12967–12978. PMLR (2021)

The Value of Knowledge: Joining Reward and Epistemic Certainty Optimisation for Anxiety-Sensitive Planning

Linda Gutsche[1(✉)] and Loïs Vanhée[2]

[1] DI ENS, École normale supérieure, Université PSL, CNRS, 75005 Paris, France
linda.gutsche@ens.psl.eu
[2] Department of Computing Science, Umeå University, 907 36 Umeå, Sweden
lois.vanhee@umu.se

Abstract. Anxiety is one of the most basic emotional states and also the most common disorder. AI agents however are typically focused on maximising performance, concentrating on expected values and disregarding the degree of exposure to uncertainty. This paper introduces a formalism derived from Partially Observable Markov Decision Processes (POMDPs) to give the first model based on cognitive psychology of the anxiety induced by epistemic uncertainty (i.e. the lack of precision of knowledge about the current state of the world). An algorithm to generate policies balancing reward maximisation and anxiety reduction is given. It is then used on a classical example to demonstrate how this can lead in some cases to a dramatic reduction of epistemic uncertainty for nearly no cost and thus a more human-friendly reward optimisation. The empirical validation shows results reminiscent of behaviours that cognitive psychology identifies as coping mechanisms to anxiety.

Keywords: Anxiety theories from psychology · Computational model of epistemic uncertainty · POMDPs

1 Introduction

Anxiety is a future-oriented emotional state experienced by all humans. It prepares them for threats, to the risk of their goals being thwarted, by increasing their ability to predict danger and react to it [6]. However, in the context of our modern society, it often results in ill-adapted responses, disturbing one's cognition, emotional regulation, or behaviour. Anxiety is the most common disorder. In 2019, 301 million people, among which 58 million adolescents and children, were suffering from anxiety disorder [14], and it has been estimated that this number increased by 26% during 2020 because of the COVID-19 pandemic [21]. Yet, in the field of Artificial Intelligence (AI), anxiety is scarcely taken into account.

Problems involving partially observable environments require agents to deliberate in situations of *epistemic uncertainty*, i.e. with only a partial knowledge of the current world-state. Partially Observable Markov Decision Processes (POMDPs) [16] are one of the most common framework for modeling and

© The Author(s), under exclusive license to Springer Nature Switzerland AG 2024
F. Amigoni and A. Sinha (Eds.): AAMAS 2023 Workshops, LNAI 14456, pp. 30–42, 2024.
https://doi.org/10.1007/978-3-031-56255-6_2

planning within such environments and have seen a wide array of applications [9], including structural inspection and maintenance problems [23], autonomous robots navigation [15], or medical diagnosis [31]. POMDP planning mostly revolves around *reward-maximisation*, i.e. maximising rewards that are obtained when reaching particular states or performing a particular action.

However, such an exclusive focus on rewards can cause systems to become blind to other critical aspects involved in acting in partially-observable environments. Thinking of anxiety as associating a negative value to uncertainty, having AI agents take anxiety into account would encourage them to create and maintain a sufficient degree of epistemic certainty while conducting operations. This carries valuable properties that is hidden to the pure system-oriented thinking. These properties include: 1) eased interactions with humans, who generally prefer environments and systems offering greater certainty and can more easily interpret information-driven deliberation [13,22]; and 2) greater flexibility for altering system use (e.g. changing goals, extending horizon).

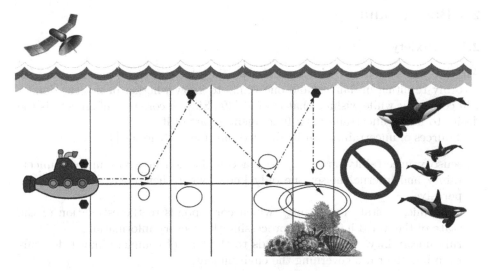

Fig. 1. Illustration of a reward maximisation approach (full line) and of our anxiety-sensitive approach balancing reward and epistemic certainty (dotted line). Black dots indicate certainty, and the greater the empty ovals, the least certain the submarine is about its location.

As an example illustrated in Fig. 1, a reward-driven policy for a submarine will likely seek to reach the goal following a straight line relying on inertial navigation sensors, saving the time needed to surface to cross-check its position via GPS. As a side effect, the system becomes increasingly uncertain about its position, and operators would thus have a hard time regaining control over such a system due to the system having no information to share. Furthermore, even if no human intervention happens, lack of precise knowledge can be anxiety inducing in those who oversees the submarine movements. In systems where humans are involved, gathering epistemic certainty is inherently valuable and should be at par and best streamlined with reward-optimisation.

Taking an interdisciplinary approach, this paper is about creating a computational model, based on psychology theories, of the anxiety induced by epistemic uncertainty. The goal is to have a model to estimate the psychological strain of strategies and suggest alternative approaches that can help alleviate anxiety. We 1) introduce the $R\rho$-POMDP formalism, a variant of the extension ρ-POMDP of POMDP that integrates epistemic certainty as a form of reward, 2) define a model of anxiety on $R\rho$-POMDPs, 3) give a Monte-Carlo-based algorithm for computing anxiety-sensitive $R\rho$-POMDP policies, 4) empirically validate that the algorithm generates policies that trade-off reward-optimisation and anxiety-minimisation, and that in a way that evokes anxiety-induced behaviours described by fundamental psychology, 5) conclude by a discussion on the perspectives opened by our model.

2   Background

2.1   Anxiety

Anxiety can be defined as a mental state in which a future event is seen as possibly resulting in danger, and one wants to know whether or not the danger will come true while wishing that it wont [19]. Since a core part of anxiety is the desire to know, uncertainty is a fundamental part of it.

Sources of uncertainty can be divided into four categories [3]:

- sensory uncertainty, arousing from a combination of environmental uncertainty and internal noise: one only knows what their noisy sensory inputs perceive,
- epistemic, or state uncertainty, which corresponds to the estimation of the state of the world inferred by processing the sensory information,
- rule uncertainty, which corresponds to the lack of complete knowledge surrounding the rules governing the environment,
- and outcome uncertainty, which corresponds to the uncertainty surrounding which one of the possible results will happen

A model for anxiety based on outcome uncertainty [28], using the framework of Markov Decision Processes, has already been established. In this paper, we focus on anxiety stemming from state uncertainty. Situations where state anxiety can be experienced include medical diagnosis, where the patient worries about the state of their body and has to chose whether or not to undergo further tests or procedures. Similarly, in infrastructure maintenance, the facility manager has to decide upon inspections and maintenance interventions. Another example is surveillance, where the agent may have to decide on which camera image to focus on to check for the possible presence of a threat.

2.2   Partially Observable Markov Decision Processes

Partially Observable Markov Decision Processes (POMDPs) allow modeling environments in which the consequences of actions are uncertain, the current state is uncertain, and a partial certainty can be acquired through observations, which are gathered in response to agent actions in relation to the current world-state [16]. A **POMDP** is formally defined as a tuple $(\mathcal{S}, \mathcal{A}, \mathcal{O}, P, Z, R, b_0, \gamma, T)$ where:

- $\mathcal{S}$ represents a set of states; $\mathcal{A}$ a set of actions, and $\mathcal{O}$ a set of *observations*;
- $P : \mathcal{S} \times \mathcal{A} \times \mathcal{S} \rightarrow [0, 1]$ represents a *transition function*, where $P(s, a, s')$ is the probability of reaching s' when applying action a from state s;
- $Z : \mathcal{S} \times \mathcal{A} \times \mathcal{S} \times \mathcal{O} \rightarrow [0, 1]$ is an *observation probability function* where $Z(s, a, s', o)$ is the probability of observing o when applying a from s and reaching s';
- $R : \mathcal{S} \times \mathcal{A} \rightarrow \mathbb{R}$ represents a *reward function*, where $R(s, a)$ is the reward for playing a from s (note that the agent only sees observations and is not informed of the reward it receives);
- $b_0 \in P(\mathcal{S})$ represents a probability distribution over states representing the *initial belief*;
- $\gamma \in [0, 1)$ is a discount factor that represents the diminishing of the importance of future rewards
- $T \in \mathbb{N} \cup \{+\infty\}$ is the *horizon* of the POMDP, corresponding to the number of actions that can be taken.

Beliefs [25] constitute a classic approach for modeling knowledge about the current state. They represent the possible current states as a probability distribution over $\mathcal{S}$: $b(s)$ corresponds to the probability of being in state s.

Beliefs can be updated to capture the evolution of the probability of being in a given state as follows: given a belief b, a performed action a, and a received observed observation o, the subsequent belief is represented as:

$$b'(s') = B(b, a, o)(s') = \frac{\displaystyle\sum_{s \in \mathcal{S}} Z(s, a, s', o) P(s, a, s') b(s)}{\displaystyle\sum_{s \in \mathcal{S}} \sum_{s'' \in \mathcal{S}} Z(s, a, s'', o) P(s, a, s'') b(s)}$$

The definition of rewards can be expanded for beliefs with

$$r(b, a) = \sum_{s \in \mathcal{S}} R(s, a) b(s) \tag{1}$$

A policy $\pi : \mathcal{B} \rightarrow \mathcal{A}$ represents strategies agents can follow, defining the action to apply given a particular belief b. The expected reward, or *value*, of a policy π for a horizon $h \in \mathbb{N}$ is defined as:

$$V_h^{\pi}(b) = r(b, \pi(b)) + \gamma \cdot \sum_{s \in \mathcal{S}} b(s) \cdot \sum_{(s', o) \in \mathcal{S} \times \Omega} w(s, \pi(b), s', o) \cdot V_{h-1}^{\pi}(B(b, \pi(b), o))$$

if $h > 0$, and $V_h^\pi(b) = 0$ otherwise, where

$$w(s, \pi(b), s', o) = P(s, \pi(b), s') \cdot Z(s, \pi(b), s', o)$$

Classic POMDP papers focus on computing a policy $\pi_R^* \in \text{argmax}_\pi \ V_T^\pi(b^0)$. To distinguish between reward maximisation and epistemic certainty, we refer to classic POMDPs as R-POMDPs (in reference to the reward R). We use the term R-value to refer to the value of a policy π, and denote it as $RV_T^\pi(b_0)$.

2.3   Epistemic Partially Observable Markov Decision Processes

ρ-POMDPs [1], also named epistemic POMDPs, constitute a variant of POMDPs in which the reward of the agent is tied to its belief instead of its current state and action. Formally, a ρ-**POMDP** is defined as a tuple $(\mathcal{S}, \mathcal{A}, \mathcal{O}, P, Z, \rho, b_0, \gamma, T)$ where $\mathcal{S}, \mathcal{A}, \mathcal{O}, P, Z, b_0, \gamma$, and T are defined similarly to R-POMDPs and $\rho :$ $\mathcal{B} \times \mathcal{A} \to \mathbb{R}$ replaces the classic reward function R by associating rewards to beliefs and actions instead of state-action pairs.

The value of a policy π in a ρ-POMDP, referred to in this paper as the ρ-value of π, is defined as:

$$\rho V_h^\pi(b) = \rho(b, \pi(b)) + \gamma \cdot \sum_{s \in \mathcal{S}} b(s) \cdot \sum_{(s', o) \in \mathcal{S} \times \Omega} w(s, \pi(b), s', o) \cdot \rho V_{h-1}^\pi(B(b, \pi(b), o))$$

if $h > 0$, else $\rho V_h^\pi(b) = 0$. Classic ρ-POMDP articles focus on computing a policy π_ρ^* that maximises $\rho V_T^{\pi_\rho^*}(b^0)$.

3 $R\rho$-POMDP Formalism

$R\rho$-POMDPs are defined as the integration between classic R-POMDPs (based on a reward R) and epistemic ρ-POMDPs (based on an epistemic reward ρ).

Even though the reward R can be integrated in ρ through the formula (2), we here separate the two, to have R stand as an external reward and ρ as an epistemic reward, so as to be able to use the two separately (which is necessary for the anxiety model).

Definition 1. *$R\rho$-POMDPs are defined as a tuple $(\mathcal{S}, \mathcal{A}, \mathcal{O}, P, Z, R, \rho, W, \mu, b_0, \gamma, T)$ where :*

- *$(\mathcal{S}, \mathcal{A}, \mathcal{O}, P, Z, R, b_0, \gamma, T)$ defines a classic POMDP,*
- *$\rho : \mathcal{B} \times \mathcal{A} \to \mathbb{R}$ (with $\mathcal{B}$ the set of probability distributions on $\mathcal{S}$) stands for an epistemic reward function defined over the set of beliefs,*
- *$W \in [0; 1]$ is a **weighting factor** that weights the relative importance of epistemic belief-based rewards when compared to external state-based rewards*
- *and $\mu \in \mathbb{R}_+$ is a **magnitude correction factor** that allows scaling values associated to epistemic certainty with rewards.*

To obtain a $R\rho$-POMDP from a R-POMDP, one only needs to incorporate ρ and the combination factors μ and W.

Definition 2. *The Rρ-value of a policy* π *is* $R\rho V_T^\pi(b_0) = (1-W) \cdot RV_T^\pi(b_0) + W \cdot \mu \cdot \rho V_T^\pi(b_0)$.

Optimising a $R\rho$-POMDP corresponds to looking for a policy π^* that maximises $R\rho V_T^{\pi^*}(b_0)$

Informally, a $R\rho$-POMDP builds over the POMDP world model (i.e. $\mathcal{S}$, $\mathcal{A}$, $\mathcal{O}$, P, Z, b_0 and T) and incorporates both the classic R-POMDP rewards and the ρ-POMDP epistemic certainty rewards.

4 Anxiety Models

We now define a model of anxiety based on $R\rho$-POMDPs. The core idea is that anxiety can be simplified as an *anticipation* of the uncertainty that may be encountered in the future.

Definition 3. *The **instant uncertainty** corresponding to some belief b is*

$$H(b) = \sum_{s \in \mathcal{S}} b(s) log_2(b(s))$$

H here above is the Shannon entropy [29], a famous uncertainty metric from the field of physics. It is a classic estimator for quantifying information. The more b is uniformly distributed, the higher $H(b)$.

Definition 4. *The **instant anxiety** for the belief b, policy π, discount γ, and horizon h, is the expected cumulative uncertainty. In a practical form, it can be represented as* $X(\pi, b) = \rho V_h^\pi(b)$ *where* $\rho(b, \cdot) := H(b)$

And thus, in the same way that the R-value of a policy is its expected cumulative reward,

Definition 5. *The **expected cumulative anxiety** felt under horizon h, initial belief b, and for the policy π, is the X-value associated with* $X(\pi, b) = \rho V_h^\pi(b)$ *obtained from* $\rho(b, \cdot) := H(b)$:

$$XV_h^\pi(b) = \begin{cases} \rho V_0^\pi(b) = \rho(b, \pi(b)) & if\ h = 0 \\ \rho V_h^\pi(b) + \gamma \cdot \sum_{s \in \mathcal{S}} b(s) \cdot \sum_{(s',o) \in \mathcal{S} \times \Omega} w(s, \pi(b), s', o) \cdot XV_{h-1}^\pi(B(b, \pi(b), o)) \end{cases}$$

Here, balancing the optimisation of cumulative reward and the minimisation of cumulative exposure to anxiety thus corresponds to finding the policy π that maximises

$$RXV_T^\pi(b_0) = (1 - W) \cdot RV_T^\pi(b_0) + W \cdot \mu \cdot (-XV_T^\pi(b_0)) \tag{2}$$

In this paper, we take $\mu = (\max_{s,s',a,a'}(r(s,a) - r(s',a')))/(log_2(m))$ where m is the number of states of the $R\rho$-POMDP.

Definition 6. *Policies that balance reward optimisation and exposure to anxiety are called in this paper **anxiety-sensitive policies**.*

5   Anxiety-Sensitive Planning Algorithm

SEARCH(h) :

1: $k = 0$
2: **while** $k < nb\_sim$ **do**
3:     $k \leftarrow k + 1$
4:     **if** $h = empty$ **then**
5:        $s, \beta \overset{n+1}{\sim} \mathcal{I}$
6:     **else**
7:        $s, \beta \overset{n+1}{\sim} B(h)$
8:     **end if**
9:     SIMULATE($s, \beta, h, 0$)
10: **end while**
11: **return** $\text{argmax}_a RX(V(ha))$

PF($\beta, a, o, s', w_{s'}$) :

1: $\beta' \leftarrow$ all weights to 0
2: $\beta'(s') \mathrel{+}= w_{s'}$
3: **for** $i \in [\![1; n]\!]$ **do**
4:     $\tilde{s} \sim \beta$
5:     $\tilde{s}' \sim \mathcal{G}(\tilde{s}, a)$
6:     $\beta'(\tilde{s}') \mathrel{+}= \mathbb{P}(o|\tilde{s}, a, \tilde{s}')$
7: **end for**
8: **return** β'

ROLLOUT($s, \beta, h, depth$) :

1: **if** $depth \geq T$ **then**
2:     **return**$(0, \rho(\beta), \rho(\beta))$
3: **end if**
4: $a \sim \pi_{rollout}(h, \cdot)$
5: $(s', o, r) \sim \mathcal{G}(s, a)$
6: $\beta' \leftarrow$ PF($\beta, a, o, s', \mathbb{P}(o|s, a, s')$)
7: **return** $(r, \rho(\beta, a)) \mathbin{\widehat{+}} \gamma \cdot$ ROLLOUT
    $(s', \beta', hao, depth + 1)$

SIMULATE($s, \beta, h, depth$) :

1: **if** $depth \geq T$ **then**
2:     **return** $(0, \rho(\beta), \rho(\beta))$
3: **end if**
4: **if** $h \notin Tree$ **then**
5:     **for all** $a \in \mathcal{A}$ **do**
6:        $Tree(ha) \leftarrow (N_{init}(ha),$
       $V_{init}(ha), \emptyset)$
7:     **end for**
8:     **return** ROLLOUT($s, \beta, h, depth$)
9: **end if**
10: $a \leftarrow \text{argmax}_b \; (RX(V(hb))$
    $+c\sqrt{\frac{\log N(h)}{N(hb)}}\,)$
11: $(s', o, r) \sim \mathcal{G}(s, a)$
12: $\beta' \leftarrow$ PF($\beta, a, o, s', \mathbb{P}(o|s, a, s')$)
13: $B(h) \leftarrow B(h) \cup \beta$
14: $(v_R, v_\rho, v_X) \leftarrow (r, \rho(B(h)), 0)$
    $\mathbin{\widehat{+}} \gamma \cdot$ SIMULATE($s', \beta', hao, depth + 1$)
15: $v_X \leftarrow v_\rho + v_X$
16: $N(h) \leftarrow N(h) + 1$
17: $N(ha) \leftarrow N(ha) + 1$
18: $V(ha) \leftarrow V(ha)$
    $\mathbin{\widehat{+}} \frac{(v_R, v_\rho, v_X) \mathbin{\widehat{-}} V(ha)}{N(ha)}$
19: **return** (v_R, v_ρ, v_X)

Fig. 2. The $R\rho X$-POMCP(β) algorithm

We propose an algorithm called $R\rho X$-POMCP(β) for finding anxiety-sensitive policies. It is based on Partially Observable Monte-Carlo Planning algorithms, which are used for solving POMDPs [26] and ρ-POMDPs [27]. They are online algorithms that perform a Monte-Carlo tree search from the current belief state and simultaneously update the agent's belief along the tree.

In a nutshell, our algorithm, given in Fig. 2, expands the simulation-based ρ-POMCP(β) [27] by integrating

- R-value estimations as done by POMCP for POMDPs,
- ρ-value estimations based on the computation of the belief b at this point of the simulation, as done by ρ-POMCP(β) for ρ-POMDPs,
- and X-value estimations based on the ρ-value ones.

The Monte-Carlo tree-searched based algorithm searches for the next action to perform by building a search tree of histories through simulations of runs. The SEARCH procedure starts each simulations at depth 0 of the tree, picking a starting state according to the current belief, and SIMULATE and ROLLOUT execute the steps of the simulated runs. When a node is visited for the first time, ROLLOUT chooses the action to perform following a $\pi_{rollout}$ policy. When a node has already been visited, SIMULATE estimates the best action to take based on the approximated values stored in the node.

To approximate the belief corresponding to a history, a particle bag $B(h)$ is associated to each node. It stores weights for each state attained when achieving the given history. In practice, after selecting a and sampling s' and o from the generative model $\mathcal{G}$, $B(hao)$ is updated with a small particle bag β' obtained through a particle filtering process PF. PF not only takes s' with a weight $w_{s'} = \mathbb{P}(o|s, a, s')$, but also generates n other particles: states sampled with their respective weights consistent with observation o. This serves to improve belief estimations, and thus also ρ-value and X-value estimations.

The estimated belief can be obtained from a particle bag by normalising its weights.

Here,

- $RX((V_R, V_\rho, V_X)) := (1 - W_\rho) \cdot V_R + W_\rho \cdot \mu \cdot \rho V_X$
- $\mathcal{I}$ is the initial belief,
- $B(h)$ is the cumulative bag of particles for history h
- $\widehat{+}$ and $\widehat{-}$ are term-by-term tuples operations
- $s, \beta \overset{n+1}{\sim} B$: sampling s as well as n other states from B to complete β
- and $\mathcal{G}(s, a)$ is a procedure applying the action a to the state s and returning the obtained new state, observation and reward s', o, r

6   Experiments

6.1   The Bridge Inspection Problem

In this problem, introduced in [12], the agent is in charge of maintaining a bridge that degrades over time. Each round, the agent must decide on a pair of two actions, consisting of one *inspection* action and one *maintenance* action, resulting in 12 possible actions in total. The bridge can be in one of the following five states: *less*, *between*, *betweentwo*, *more*, and *failed* (s_{failed}), with s_{failed} being a pit state highly punished by R. The three inspection actions are: *no-inspect*, *visual-inspect*, and *ut-inspect*. The first one does not give any information, the visual inspection can only estimate if the bridge is in good, fair, or poor condition, and is therefore less precise than the nearly always-accurate ultrasonic inspection. The four maintenance actions are: *no-repair*, *clean-paint*, *paint-strengthen*, and *structural-repair*. The structural repair is the only repair that is deep enough to bring the bridge out of the *failed* state. The more superficial the repair or inspection, the less costly it is. Similarly, the better the state of the bridge, the less it costs to repair it. Furthermore, extra costs are incurred when the bridge is in poor condition.

6.2   Setup

The models on which we tested our algorithm were downloaded from Cassandra's pomdp.org website[1]. The values of Fig. 3 correspond to the averaged results of 50 runs of the $R\rho X$-POMCP(β) algorithm trained on 20000 simulations. We measured instant rewards and expected cumulative rewards, instant anxiety and expected cumulative anxiety, as well as the evolution of uncertainty during each run. The discount γ was set to 1, and the experiments were conducted with a fixed horizon $T = 10$.

6.3   Results

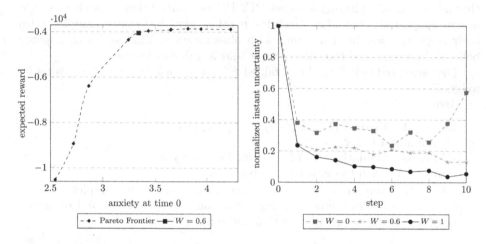

Fig. 3. Results in the Bridge Inspection environment. The graph on the left displays the Pareto frontier showing the compromises between expected cumulative reward and anxiety at time 0 for W taking 11 different values between 0 (pure expected total reward optimisation) and 1 (pure expected cumulative anxiety reduction). The graph on the right shows the mean evolution of uncertainty along 10-step runs for three different values of W. In both graphs, uncertainty was normalised to have 1 correspond to a lack of any information. The value $W = 0.6$ was selected as a good trade-off as any further efforts towards reducing anxiety would entail a steep decline in expected reward.

Testing our algorithm on POMDP environments such as the bridge example showed that anxiety-sensitive approaches can successfully reduce exposure to anxiety as measured by our model. The more instrumental alignment there is between reward optimisation and uncertainty reduction in the environment, the more exposition to anxiety can be reduced at nearly no external reward cost.

In the bridge experiment, the results of which are presented here, a rise in uncertainty can be observed at the end of the $W = 0$ runs. This rise is due

---

[1] https://www.pomdp.org/examples/ [24].

to the fact that the state of the bridge no longer needs to be monitored after the tenth step, and the purely external-reward-oriented agent is thus no longer interested in its state. However, in terms of anxiety, this rise is however very costly. It is anticipated at any time step, and thus contributes to all instant anxiety experienced throughout the run, resulting in a dramatic cumulative cost.

Finally, our results showed promising links with cognitive psychology theories, in that agents overly sensitive to anxiety seemed to reproduce pathological behaviours associated with it. On the famous tiger example [10], where the agent stands in front of two doors, one hiding a tiger and the other a large reward, and only three actions can be taken (opening one door or the other, or listening), the $W = 1$ agent ended up only listening. This behaviour is reminiscent of task-avoidance, a common behavioural response to anxiety [5]. On the bridge example, the $W = 1$ agent carefully examined the deteriorating bridge and then left it in its *failed* state, which evokes pessimistic solutions that resolve anxiety by making failure more predictable [19].

7   Discussion and Further Works

The anxiety model presented here is built on uncertainty about the entire state. In some cases, one may want to focus solely on the danger level or other core features of a state. The model can be easily adapted by changing the probability distribution to which the Shannon entropy measure is applied. Another potential adjustment would be to include a measure of the entropy of the distribution of possible future beliefs. Currently, the model of anxiety is based only on the sum of future epistemic uncertainty and does not account for the limited predictability of which possible chain of events the chosen strategy will lead to.

Additional experiments would be necessary to fully understand the potential and limitations of the model. It would be particularly interesting to devise experiments in which both human and AI agents can participate to compare their behaviours. Finally, further possible research avenues include combining this model of epistemic anxiety with the preexisting model of outcome uncertainty-induced anxiety [28]. Another interesting combination would be to join risk-sensitivity to epistemic-anxiety-sensitivity, as anxiety is concerned not only with predictability but also controllability of the environment [19].

8   Concluding Remarks

This paper deals with combining cognitive psychology and artificial intelligence to model the anxiety induced by the absence of epistemic certainty. It provides a way to estimate the level of anxiety associated with a strategy and helps in finding alternative approaches to mitigate anxiety by 1) introducing the $R\rho$-POMDP formalism that separates external rewards and epistemic rewards to allow defining tradeoffs between different exploitations of the two; 2) suggesting a model of anxiety; 3) giving the $R\rho X$-POMCP(β) algorithm, a Monte-Carlo-based algorithm for computing anxiety-sensitive POMDP policies that trade off reward maximisation and anxiety reduction; and 4) beginning experiments to validate the proposed model.

One potential application of this approach is to enable AI agents that assist humans in tasks or decisions to generate policies with lower exposure to anxiety, instead of solely maximising rewards. More generally, anxiety-sensitive policies are particularly suited when humans are involved as they drive the system towards increasing its level of information, thus increasing the opportunities for humans to take over should a special situation be encountered, generating policies that are closer to human approaches to problem-solving [13,22]. Our model paves the way for dealing with challenging environments that require optimising efficacy and information mastery, such as disaster recovery missions [7,11,30], exploration missions [4,20], and high-risk control [2]. As an important benefit, any existing POMDP-based application can be expanded with sensitivity to epistemic uncertainty-induced anxiety as $R\rho$-POMDP directly extends the classic POMDP formalism.

Furthermore, $R\rho$-POMDPs and the epistemic anxiety model open up exciting perspectives on the general topic of planning under uncertainty. They provide an additional approach to the definition and treatment of uncertainty, beyond existing definitions such as rewards, average knowledge, final knowledge, risks, outcome anxiety, hope [1,8,16–18,28] and offer solutions for achieving pragmatic compromises rather than extreme outcomes.

Finally, comparing anxiety models with actual human behaviour can be a valuable tool for gaining insight into the latter.

References

1. Araya, M., Buffet, O., Thomas, V., Charpillet, F.: A pomdp extension with belief-dependent rewards. In: Advances in Neural Information Processing Systems, vol. 23 (2010)
2. Ayer, T., Alagoz, O., Stout, N.K.: Or forum-a POMDP approach to personalize mammography screening decisions. Oper. Res. **60**(5), 1019–1034 (2012)
3. Bach, D.R., Dolan, R.J.: Knowing how much you don't know: a neural organization of uncertainty estimates. Nat. Rev. Neurosci. **13**, 572–586 (2012)
4. Balaban, E., Arnon, T., Shirley, M.H., Brisson, S.F., Gao, A.: A system health aware pomdp framework for planetary rover traverse evaluation and refinement. In: 2018 AIAA Information Systems-AIAA Infotech @ Aerospace (2018)
5. Barlow, D.: Anxiety and Its Disorders: The Nature and Treatment of Anxiety and Panic. Anxiety and Its Disorders: The Nature and Treatment of Anxiety and Panic, Guilford Publications (2004)
6. Beck, A.T., Clark, D.A.: An information processing model of anxiety: automatic and strategic processes. Behav. Res. Ther. **35**(1), 49–58 (1997)
7. Bravo, R.Z.B., Leiras, A., Cyrino Oliveira, F.L.: The use of UAVs in humanitarian relief: an application of POMDP-based methodology for finding victims. Prod. Oper. Manage. **28**(2), 421–440 (2019)
8. Broekens, J., Jacobs, E., Jonker, C.M.: A reinforcement learning model of joy, distress, hope and fear. Connect. Sci. **27**(3), 215–233 (2015)

9. Cassandra, A.R.: A survey of POMDP applications. In: Working Notes of AAAI Fall Symposium on Planning with Partially Observable Markov Decision Processes (1998)
10. Cassandra, A.R., Kaelbling, L.P., Littman, M.L.: Acting optimally in partially observable stochastic domains. In: Proceedings of the Twelfth AAAI National Conference on Artificial Intelligence, AAAI 1994, pp. 1023–1028. AAAI Press (1994)
11. Çelik, M., Ergun, O., Keskinocak, P.: The post-disaster debris clearance problem under incomplete information. Oper. Res. **63**(1), 65–85 (2015)
12. Corotis, R., Ellis, H., Jiang, M.: Modeling of risk-based inspection, maintenance and life-cycle cost with partially observable markov decision processes. Struct. Infrastruct. Eng. **1**, 75–84 (2005)
13. Duke, K.E., Goldsmith, K., Amir, O.: Is the preference for certainty always so certain? J. Assoc. Consum. Res. **3**(1), 63–80 (2018)
14. Institute of Health Metrics and Evaluation, Global Health Data Exchange. www.vizhub.healthdata.org/gbd-results/, Accessed 14 Mai 2022
15. Ibekwe, H., Kamrani, A.: Navigation for autonomous robots in partially observable facilities. In: 2014 World Automation Congress, pp. 1–5 (2014)
16. Kaelbling, L.P., Littman, M.L., Cassandra, A.R.: Planning and acting in partially observable stochastic domains. Artif. Intell. **101**(1–2), 99–134 (1998)
17. Mafi, N., Abtahi, F., Fasel, I.: Information theoretic reward shaping for curiosity driven learning in POMDPs. In: 2011 IEEE International Conference on Development and Learning (ICDL), vol. 2, pp. 1–7 (2011)
18. Marecki, J., Varakantham, P.: Risk-sensitive planning in partially observable environments. In: Proceedings of the 9th International Conference on Autonomous Agents and Multiagent Systems, AAMAS. 2, pp. 1357–1368 (2010)
19. Miceli, M., Castelfranchi, C.: Anxiety as an "epistemic" emotion: an uncertainty theory of anxiety. Anxiety Stress Coping **18**, 291–319 (2005)
20. Niroui, F., Sprenger, B., Nejat, G.: Robot exploration in unknown cluttered environments when dealing with uncertainty. In: 2017 IEEE International Symposium on Robotics and Intelligent Sensors (IRIS), pp. 224–229. IEEE (2017)
21. Geneva: World Health Organization: Mental health and covid-19: Early evidence of the pandemic's impact (2022)
22. Osmanağaoğlu, N., Creswell, C., Dodd, H.F.: Intolerance of uncertainty, anxiety, and worry in children and adolescents: a meta-analysis. J. Affect. Disord. **225**, 80–90 (2018)
23. Papakonstantinou, K., Shinozuka, M.: Planning structural inspection and maintenance policies via dynamic programming and Markov processes. Part II: POMDP implementation. Reliab. Eng. Syst. Saf. **130**, 214–224 (2014)
24. pomdp.org: Examples. www.pomdp.org/examples/, Accessed 26 Jun 2022
25. Roy, N., Gordon, G., Thrun, S.: Finding approximate POMDP solutions through belief compression. J. Artif. Intell. Res. **23**, 1–40 (2005)
26. Silver, D., Veness, J.: Monte-carlo planning in large pomdps. In: Lafferty, J., Williams, C., Shawe-Taylor, J., Zemel, R., Culotta, A. (eds.) Advances in Neural Information Processing Systems, vol. 23. Curran Associates, Inc. (2010)
27. Thomas, V., Hutin, G., Buffet, O.: Monte Carlo information- oriented planning (2021). www.arxiv.org/abs/2103.11345
28. Vanhée, L., Jeanpierre, L., Mouaddib, A.I.: Anxiety-sensitive planning: from formal foundations to algorithms and applications. In: Proceedings of the International Conference on Automated Planning and Scheduling, vol. 32, pp. 730–740 (2022)
29. Wang, Q.A.: Probability distribution and entropy as a measure of uncertainty. J. Phys. A: Math. Theor. **41**(6), 065004 (2008)

30. Wu, F., Ramchurn, S.D., Chen, X.: Coordinating human-UAV teams in disaster response. In: Proceedings of the 25th International Joint Conference on Artificial Intelligence (IJCAI), pp. 524–530 (2016)
31. Zhang, W., Wang, H.: Diagnostic policies optimization for chronic diseases based on POMDP model. In: Healthcare (2022)

Learning Reward Machines
in Cooperative Multi-agent Tasks

Leo Ardon$^{(\boxtimes)}$ ⓘD, Daniel Furelos-Blanco ⓘD, and Alessandra Russo ⓘD

Imperial College London, London, UK
{leo.ardon19,d.furelos-blanco18,a.russo}@imperial.ac.uk

Abstract. This paper presents a novel approach to Multi-Agent Reinforcement Learning (MARL) that combines cooperative task decomposition with the learning of *Reward Machines (RMs)* encoding the structure of the sub-tasks. The proposed method helps deal with the non-Markovian nature of the rewards in partially observable environments and improves the interpretability of the learnt policies required to complete a cooperative task. The RMs associated with the sub-tasks are learnt in a decentralised manner and then used to guide the behaviour of each agent in a team acting towards a common goal. By doing so, the complexity of a cooperative multi-agent problem is reduced, allowing for more effective learning. The results suggest that our approach is a promising direction for future research in cooperative MARL, especially in complex and partially observable environments.

Keywords: Multi-Agent Reinforcement Learning · Reward Machines · Neuro-Symbolic · Symbolic Machine Learning

1 Introduction

With impressive advances in the past decade [12,20,24], the Reinforcement Learning (RL) paradigm [26] appears as a promising avenue in the quest to have machines learn autonomously how to achieve a goal. Originally evaluated in the context of a single agent interacting with the rest of the world, researchers have since then extended the approach to the multi-agent setting (MARL) [2,23,25], where multiple autonomous entities learn in the same environment. Multiple agents learning concurrently brings a set of challenges, such as partial observability, non-stationarity, and scalability [1,3]; but, it also represents a more realistic setting of the problem to solve. Like in many real-life scenarios, the actions of one individual often affect the way others act and should therefore be considered in the learning process.

A sub-field of MARL called Cooperative Multi-Agent Reinforcement Learning (CMARL) focuses on learning policies for multiple agents that must coordinate their actions to achieve a shared objective. These tasks can be tackled using 'divide and conquer' strategies; that is, by decomposing them into smaller and simpler tasks assigned to different agents acting in parallel in the environment.

F. Amigoni and A. Sinha (Eds.): AAMAS 2023 Workshops, LNAI 14456, pp. 43–59, 2024.
https://doi.org/10.1007/978-3-031-56255-6_3

The problem of learning an optimal sequence of actions now also includes the need to strategically divide the task among agents that must learn to coordinate and communicate in order to find the optimal way to cooperatively accomplish the goal.

An emergent field of research in the RL community exploits finite-state machines to encode the structure of the reward function; ergo, the structure of the task at hand. This new construct called *Reward Machine* (RM) [27] can model non-Markovian reward and provides a symbolic interpretation of the stages required to complete a task. While RMs can be handcrafted, it is often impractical to do so. Hence, learning RMs has been the object of several works in recent years [6,13,14,17,29,30] as a way to reduce human input. However, the aforementioned challenges in the multi-agent setting hinder the learning of an RM encoding the global task. With multiple agents at play, the number of states of the global RM grows exponentially, making its learning extremely challenging.

In this paper, we argue that learning RMs for smaller tasks obtained from an appropriate task decomposition is more efficient. The global RM can later be reconstructed through a parallel composition of the learned RMs for the sub-tasks. Inspired by previous work on using RMs to decompose and distribute a global task to a team of collaborative agents [21], we propose a method that combines task decomposition with the autonomous learning of RMs in a multi-agent setting. Our approach offers a more scalable and interpretable way to learn the structure of a complex task involving collaboration among multiple agents. We present an algorithm that learns (in parallel) the RM for each agent's sub-task and the associated RL policies that enable agents to achieve the global cooperative task. We compare our approach with (i) a method employing handcrafted RMs [21], and (ii) a decentralized Q-learning approach that does not leverage the task's structure. We empirically show in two existing environments [21] that with our method a team of agents is able to learn how to solve a collaborative task more efficiently (i.e. with fewer training iterations) than without using RMs. The policies trained with our approach also achieve similar performance than with handcrafted RMs while requiring less human input, thus demonstrating the value of the proposed technique. Finally, the RMs learnt with our algorithm are easily interpretable and accurately characterize the structure of the task at hand.

2   Background

Given a finite set $\mathcal{X}$, $\Delta(\mathcal{X})$ denotes the probability simplex over $\mathcal{X}$, $\mathcal{X}^*$ denotes (possibly empty) sequences of elements from $\mathcal{X}$, and $\mathcal{X}^+$ is a non-empty sequence. The symbols $\bot$ and $\top$ denote false and true, respectively.

2.1   Reinforcement Learning

We consider T-*episodic labeled* Markov decision processes (MDPs) [30], characterized by a tuple $\langle \mathcal{S}, s_I, \mathcal{A}, p, r, T, \gamma, \mathcal{P}, l \rangle$ consisting of a set of states $\mathcal{S}$, an initial state $s_I \in \mathcal{S}$, a set of actions $\mathcal{A}$, a transition function $p : \mathcal{S} \times \mathcal{A} \to \Delta(\mathcal{S})$, a *not* necessarily Markovian reward function $r : (\mathcal{S} \times \mathcal{A})^+ \times \mathcal{S} \to \mathbb{R}$, the time horizon T of each episode, a discount factor $\gamma \in [0, 1)$, a finite set of propositions $\mathcal{P}$, and a labeling function $l : \mathcal{S} \times \mathcal{S} \to 2^{\mathcal{P}}$.

A (state-action) *history* $h_t = \langle s_0, a_0, \ldots, s_t \rangle \in (\mathcal{S} \times \mathcal{A})^* \times \mathcal{S}$ is mapped into a *label trace* (or trace) $\lambda_t = \langle l(s_0, s_1), \ldots, l(s_{t-1}, s_t) \rangle \in (2^{\mathcal{P}})^+$ by applying the labeling function to each state transition in h_t. We assume that the reward function can be written in terms of a label trace, i.e. formally $r(h_{t+1}) = r(\lambda_{t+1}, s_{t+1})$. The goal is to find an optimal *policy* $\pi : (2^{\mathcal{P}})^+ \times \mathcal{S} \to \mathcal{A}$, mapping (traces-states) to actions, in order to maximize the expected cumulative discounted reward (or return) $R_t = \mathbb{E}_\pi \left[\sum_{k=t}^{T} \gamma^{k-t} r(\lambda_{k+1}, s_{k+1}) \right]$.

At time t, the agent keeps a trace $\lambda_t \in (2^{\mathcal{P}})^+$, and observes state $s_t \in \mathcal{S}$. The agent chooses the action to execute with its policy $a = \pi(\lambda_t, s_t) \in \mathcal{A}$, and the environment transitions to state $s_{t+1} \sim p(\cdot \mid s_t, a_t)$. As a result, the agent observes a new state s_{t+1} and a new label $\mathcal{L}_{t+1} = l(s_t, s_{t+1})$, and gets the reward $r_{t+1} = r(\lambda_{t+1}, s_{t+1})$. The trace $\lambda_{t+1} = \lambda_t \oplus \mathcal{L}_{t+1}$ is updated with the new label.

In this work, we focus on the *goal-conditioned* RL problem [18] where the agent's task is to reach a goal state as rapidly as possible. We thus distinguish between two types of traces: *goal* traces and *incomplete* traces. A trace λ_t is said to be a *goal* trace if the task's goal has been achieved before t otherwise we say that the trace is *incomplete*. Formally, the MDP includes a *termination function* $\tau : (2^{\mathcal{P}})^+ \times \mathcal{S} \to \{\bot, \top\}$ indicating that the trace λ_t at time t is an *incomplete* trace if $\tau(\lambda_t, s_t) = \bot$ or a *goal* trace if $\tau(\lambda_t, s_t) = \top$. We assume the reward is 1 for goal traces and 0 otherwise. For ease of notation, we will refer to a goal trace as $\lambda_t^{\top}$ and an incomplete trace as $\lambda_t^{\bot}$. Note that we assume the agent-environment interaction stops when the task's goal is achieved. This is without loss of generality as it is equivalent to having a final absorbing state with a null reward associated and no label returned by the labelling function l.

The single-agent framework above can be extended to the collaborative *multi-agent* setting as a Markov game. A cooperative *Markov game* of N agents is a tuple $\mathcal{G} = \langle N, \boldsymbol{\mathcal{S}}, \boldsymbol{s}_I, \boldsymbol{\mathcal{A}}, p, r, \tau, T, \gamma, \mathcal{P}, l \rangle$, where $\boldsymbol{\mathcal{S}} = \mathcal{S}_1 \times \cdots \times \mathcal{S}_N$ is the set of joint states, $\boldsymbol{s}_I \in \boldsymbol{\mathcal{S}}$ is a joint initial state, $\boldsymbol{\mathcal{A}} = \mathcal{A}_1 \times \cdots \times \mathcal{A}_N$ is the set of joint actions, $p : \boldsymbol{\mathcal{S}} \times \boldsymbol{\mathcal{A}} \to \Delta(\boldsymbol{\mathcal{S}})$ is a joint transition function, $r : (\boldsymbol{\mathcal{S}} \times \boldsymbol{\mathcal{A}})^+ \times \boldsymbol{\mathcal{S}} \to \mathbb{R}$ is the collective reward function, $\tau : (2^{\mathcal{P}})^+ \times \boldsymbol{\mathcal{S}} \to \{\bot, \top\}$ is the collective termination function, and $l : \boldsymbol{\mathcal{S}} \times \boldsymbol{\mathcal{S}} \to 2^{\mathcal{P}}$ is an event labeling function. The elements T, γ, and $\mathcal{P}$ are defined as for MDPs. Like Neary et al. [21], we assume each agent A_i's dynamics are independently governed by local transition functions $p_i : \mathcal{S}_i \times \mathcal{A}_i \to \Delta(\mathcal{S}_i)$, hence the joint transition function is $p(\boldsymbol{s}' \mid \boldsymbol{s}, \boldsymbol{a}) = \Pi_{i=1}^{N} p(s_i' \mid s_i, a_i)$ for all $\boldsymbol{s}, \boldsymbol{s}' \in \boldsymbol{\mathcal{S}}$ and $\boldsymbol{a} \in \boldsymbol{\mathcal{A}}$. The objective is to find a team policy $\pi : (2^{\mathcal{P}})^+ \times \boldsymbol{\mathcal{S}} \to \boldsymbol{\mathcal{A}}$ mapping pairs of traces and joint states to joint actions that maximises the expected cumulative collective reward.

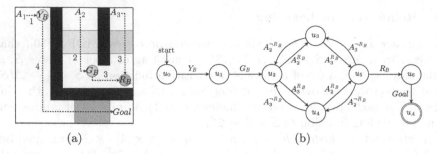

(a)   (b)

Fig. 1. Illustration of the THREEBUTTONS grid (a) and a RM modeling the task's structure (b) [21]. (Color figure online)

Example 1. We use the THREEBUTTONS task [21], illustrated in Fig. 1a, as a running example. The task consists of three agents (A_1, A_2, and A_3) that must cooperate for agent A_1 to get to the *Goal* position. To achieve this, they must open doors that prevent other agents from progressing by pushing buttons Y_B, G_B, and R_B. Agents can go UP, DOWN, LEFT, RIGHT or DO NOTHING. An agent starts pushing a button when it steps on the button's position; likewise, an agent stops pushing a button when it leaves the button's position. The button of a given colour opens the door of the same colour. Two agents are required to push R_B at once, involving synchronization. Once a door has been opened, it remains open until the end of the episode. The order in which high-level actions must be performed is shown in the figure: (1) A_1 pushes Y_B, (2) A_2 pushes G_B, (3) A_2 and A_3 push R_B, and (4) A_1 reaches *Goal*. We can formulate this problem by defining a shared reward function returning 1 when the *Goal* position is reached. The states solely consist of the position of the agents in the environment. The *proposition set* in this task is $\mathcal{P} = \{R_B, Y_B, G_B, A_2^{\neg R_B}, A_2^{R_B}, A_3^{\neg R_B}, A_3^{R_B}, Goal\}$, where (i) R_B, Y_B and G_B indicate that the red, yellow and green buttons have been pushed respectively, (ii) $A_i^{R_B}$ (resp. $A_i^{\neg R_B}$), where $i \in \{2,3\}$, indicates that agent A_i has started (resp. has stopped) pushing the red button R_B, and (iii) *Goal* indicates that the goal position has been reached. A *goal trace* example is $\langle \{\}, \{Y_B\}, \{G_B\}, \{A_2^{R_B}\}, \{A_3^{R_B}\}, \{R_B\}, \{Goal\} \rangle$.

2.2 Reward Machines

We here introduce Reward Machines, the formalism we use to express decomposable (multi-agent) tasks.

Definitions. A Reward Machine [27,28] is a finite-state machine that represents the reward function of an RL task. Formally, an RM is a tuple $M = \langle \mathcal{U}, \mathcal{P}, u_0, u_A, \delta_u, \delta_r \rangle$ where $\mathcal{U}$ is a set of states; $\mathcal{P}$ is a set of propositions; $u_0 \in \mathcal{U}$ is the initial state; $u_A \in \mathcal{U}$ is the final state; $\delta_u : \mathcal{U} \times 2^{\mathcal{P}} \to \mathcal{U}$ is a state-transition function such that $\delta_u(u, \mathcal{L})$ is the state that results from observing label $\mathcal{L} \in 2^{\mathcal{P}}$ in state $u \in \mathcal{U}$; and $\delta_r : \mathcal{U} \times \mathcal{U} \to \mathbb{R}$ is a reward-transition function such that

$\delta_r(u, u')$ is the reward obtained for transitioning from state $u \in \mathcal{U}$ to $u' \in \mathcal{U}$. We assume that (i) there are no outgoing transitions from u_A, and (ii) $\delta_r(u, u') = 1$ if $u' = u_A$ and 0 otherwise (following the reward assumption in Sect. 2.1). Given a trace $\lambda = \langle \mathcal{L}_0, \ldots, \mathcal{L}_n \rangle$, a *traversal* is a sequence of RM states $\langle v_0, \ldots, v_{n+1} \rangle$ such that $v_0 = u_0$, and $v_{i+1} = \delta_u(v_i, \mathcal{L}_i)$ for $i \in \{1, \ldots, n\}$.

Ideally, RMs are such that (i) the cross product $\mathcal{S} \times \mathcal{U}$ of their states with the MDPs' make the reward and termination functions Markovian, and (ii) traversals for goal traces end in the final state u_A, and traversals for incomplete traces do not.

Example 2. Figure 1b shows an RM for the THREEBUTTONS task. For simplicity, edges are labelled using the single proposition that triggers the transitions instead of sets. The goal trace from Example 1 is associated with the traversal $\langle u_0, u_1, u_2, u_3, u_5, u_6, u_A \rangle$.

RL Algorithm. The Q-learning for RMs (QRM) algorithm [27,28] exploits the task structure modelled through RMs. QRM learns a Q-function $q_u : \mathcal{S} \times \mathcal{A} \to \mathbb{R}$ for each state $u \in \mathcal{U}$ in the RM, which estimates the expected return from a given state-action pair. Given an experience tuple $\langle s, a, s' \rangle$, a Q-function q_u is updated according to the Bellman update as follows:

$$q_u(s, a) = q_u(s, a) + \alpha \left(\delta_r(u, u') + \gamma \max_{a'} q_{u'}(s', a') - q_u(s, a) \right),$$

where $u' = \delta_u(u, l(s, s'))$. All Q-functions (or a subset of them) of the RM are updated at each step using the same experience tuple in a counterfactual manner, by evaluating the effect of the experience tuple in all the states of the RM. QRM is guaranteed to converge to an optimal policy in the tabular case.

Multi-agent Decomposition. Neary et al. [21] recently proposed to decompose the RM for a multi-agent task into several RMs (one per agent) executed in parallel. The task decomposition into individual and independently learnable sub-tasks addresses the problem of non-stationarity inherent in the multi-agent setting. In addition, the use of RMs gives the high-level structure of the task each agent ought to solve. In this setting, each agent A_i has its own RM M_i, local state space $\mathcal{S}_i$ (e.g., it solely observes its position in the grid), and propositions $\mathcal{P}_i$ such that $\mathcal{P} = \bigcup_i \mathcal{P}_i$. Besides, instead of having a single opaque labelling function, each agent A_i employs its own labelling function $l_i : \mathcal{S}_i \times \mathcal{S}_i \to 2^{\mathcal{P}_i}$. Each labelling function is assumed to return at most one proposition per agent per timestep, and they should together output the same label as the global labelling function l.

Given a global RM M modelling the structure of the task at hand (e.g., that in Fig. 1b) and each agent's proposition set $\mathcal{P}_i$, Neary et al. [21] propose a method for deriving each agent's RM M_i by projecting M onto $\mathcal{P}_i$. We refer the reader to the paper for a description of the projection mechanism and its guarantees since we focus on learning these individual RMs instead (see Sect. 3).

Example 3. Given the local proposition sets $\mathcal{P}_1 = \{Y_B, R_B, Goal\}$, $\mathcal{P}_2 = \{Y_B, G_B, A_2^{R_B}, A_2^{\neg R_B}, R_B\}$ and $\mathcal{P}_3 = \{G_B, A_3^{R_B}, A_3^{\neg R_B}, R_B\}$, Fig. 2 shows the RMs that result from applying Neary et al.'s [21] projection algorithm.

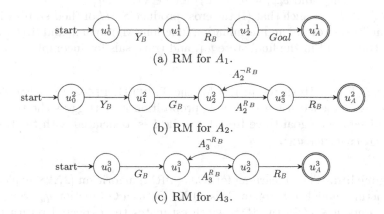

(a) RM for A_1.

(b) RM for A_2.

(c) RM for A_3.

Fig. 2. RMs for each of the agents in THREEBUTTONS [21].

Neary et al. [21] extend QRM and propose a decentralised training approach to train each agent in isolation (i.e., in the absence of their teammates) exploiting their respective RMs; crucially, the learnt policies should work when all agents interact simultaneously with the world. In both the individual and team settings, agents must synchronize whenever a shared proposition is observed (i.e., a proposition that exists in the proposition set of two or more agents). Specifically, an agent should check with all its teammates whether their labelling functions also returned the same proposition. During training, given that the agent is learning in isolation and therefore the rest of the team is not actually acting, synchronization is *simulated* with a fixed probability of occurrence.

3 Learning Reward Machine in Cooperative Multi-agent Tasks

Decomposing a task using RMs is a promising approach to solving collaborative problems where coordination and synchronization among agents are required, as described in Sect. 2.2. In well-understood problems, such as the THREEBUTTONS task, one can manually engineer the RM encoding the sequence of sub-tasks needed to achieve the goal. However, this becomes a challenge for more complex problems, where the structure of the task is unknown. In fact, the human-provided RM may hinder learning if the human intuition about the structure of the task is incorrect. To alleviate this issue, we argue for learning the RM automatically from traces collected via exploration instead of handcrafting them. In what follows, we describe two approaches to accomplish this objective.

3.1   Learn a Global Reward Machine

A naive approach to learning each agent's RM consists of two steps. First, a global RM M is learnt from traces where each label is the union of each agent's label. Different approaches [13,17,29,30] could be applied in this situation. The learnt RM is then decomposed into one per agent by projecting it onto the local proposition set $\mathcal{P}_i$ of each agent A_i [21].

Despite its simplicity, this method is prone to scale poorly. It has been observed that learning a minimal RM (i.e., an RM with the fewest states) from traces becomes significantly more difficult as the number of constituent states grows [13]. Indeed, the problem of learning a minimal finite-state machine from a set of examples is NP-complete [16]. Intuitively, the more agents interacting with the world, the larger the global RM will become, hence impacting performance and making the learning task more difficult. Moreover, having multiple agents potentially increases the number of observable propositions as well as the length of the traces; and even more so with sub-optimal policies, increasing the complexity of the problem even further.

3.2   Learn Individual Reward Machine

We propose to decompose the global task into sub-tasks, each of which can be independently solved by one of the agents. Note that even though an agent is assigned a sub-task, it does not have any information about its structure nor how to solve it. This is in fact what the proposed approach intends to tackle: learning the RM encoding the structure of the sub-task. This should be simpler than learning a global RM since the number of constituent states and propositions is smaller.

Given a set of N agents, we assume that the global task can be decomposed into N sub-task MDPs. Similarly to Neary et al. [21], we assume each agent's MDP has its own state space $\mathcal{S}_i$, action space $\mathcal{A}_i$, transition function p_i, and termination function τ_i. In addition, each agent has its own labelling function l_i; hence, the agent may only observe a subset of propositions $\mathcal{P}_i \subseteq \mathcal{P}$. Note that the termination function is particular to each agent; for instance, agent A_2 in THREEBUTTONS will complete its sub-task by pressing the green button, then the red button until it observes the signal R_B indicating that A_3 is also pressing it. Ideally, the parallel composition of the learned RMs captures the collective goal.

We propose an algorithm that *interleaves* the induction of the RMs from a collection of label traces and the learning of the associated Q-functions, akin to that by Furelos-Blanco et al. [13] for the single agent setting. The decentralised learning of the Q-functions is performed using the method by Neary et al. [21], briefly outlined in Sect. 2.2. The induction of the RMs is done using a state-of-the-art inductive logic programming system, ILASP [19], which learns the transition function of an RM as a set of logic rules from example traces observed by the agent.

Example 4. The edge connecting u_0^1 and u_1^1 in Fig. 2a corresponds to the rule $\delta(u_0^1, u_1^1, \mathtt{T}) \colon\!\!\text{-} \mathtt{prop}(Y_B, \mathtt{T})$. The $\delta(\mathtt{X}, \mathtt{Y}, \mathtt{T})$ atoms express that the transition from $\mathtt{X}$ to $\mathtt{Y}$ is satisfied at time $\mathtt{T}$, while the $\mathtt{prop}(\mathtt{P}, \mathtt{T})$ atoms indicate that proposition $\mathtt{P}$ is observed at time $\mathtt{T}$. The rule as a whole expresses that the transition from u_0^1 to u_1^1 is satisfied at time $\mathtt{T}$ if Y_B is observed at that time. The traces provided to the RM learner are expressed as sets of $\mathtt{prop}$ facts; for instance, the trace $\langle \{Y_B\}, \{R_B\}, \{Goal\} \rangle$ is mapped into $\{\mathtt{prop}(Y_B, 0). \ \mathtt{prop}(R_B, 1). \ \mathtt{prop}(Goal, 2).\}$.

Algorithm 1 shows the pseudocode describing how the reinforcement and RM learning processes are interleaved. The algorithm starts initializing for each agent A_i: the set of states $\mathcal{U}^i$ of the RM, the RM M^i itself, the associated Q-functions q^i, and the sets of example goal $(\varLambda_\top^i)$ and incomplete $(\varLambda_\perp^i)$ traces (1.1–5). Initially, each agent's RM solely consists of two unconnected states: the initial and final states (i.e., the agent loops in the initial state forever). Next, *NumEpisodes* are run for each agent. Before running any steps, each environment is reset to the initial state s_I^i, each RM M^i is reset to its initial state u_0^i, each agent's trace λ^i is emptied, and we indicate that no agents have yet completed their episodes (1.8–9). Then, while there are agents that have not yet completed their episodes (1.10), a training step is performed for each of the agents, which we describe in the following paragraphs. Note that we show a sequential implementation of the approach but it could be performed in parallel to improve performance.

The `TrainAgent` routine shows how the reinforcement learning and RM learning processes are interleaved for a given agent A_i operating in its own environment. From 1.17 to 1.25, the steps for QRM (i.e., the RL steps) are performed. The agent first selects the next action a to execute in its current state s_t given the Q-function q_{u_t} associated with the current RM state u_t. The action is then applied, and the agent observes the next state s_{t+1} and whether it has achieved its goal (1.18). The next label $\mathcal{L}_{t+1}$ is then determined (1.19) and used to (i) extend the episode trace λ_t (1.20), and (ii) obtain reward r and the next RM state u_{t+1} (1.21). The Q-function q is updated for both the RM state u_t the agent is in at time t (1.22) and in a counterfactual manner for all the other states of the RM (1.25).

The learning of the RMs occurs from 1.27 to 1.42. Remember that there are two sets of example traces for each agent: one for goal traces $\varLambda_\top$ and one for incomplete traces $\varLambda_\perp$. Crucially, the learnt RMs must be such that (i) traversals for goal traces end in the final state, and (ii) traversals for incomplete traces do not end in the final state. Given the trace λ_{t+1} and the RM state u_{t+1} at time $t+1$, the example sets are updated in three cases:

1. If λ_{t+1} is a goal trace (1.27), λ_{t+1} is added to $\varLambda_\top$. We consider as incomplete all sub-traces of $\lambda_\top$: $\{\langle \mathcal{L}_0, \ldots, \mathcal{L}_k \rangle; \forall k \in \{0, \ldots, |\lambda_\top| - 1\}\}$ (see Example 5). Note that if a sub-trace was a goal trace, the episode would have ended before. This optimization enables capturing incomplete traces faster; that is, it prevents waiting for them to appear as counterexamples (see next case).

Algorithm 1: Multi-agent QRM with RM Learning

---

1 **for** $i = 1$ **to** N **do**
2 $\quad$ $\mathcal{U}^i \leftarrow \{u_0^i, u_A^i\}$
3 $\quad$ $M^i \leftarrow$ InitializeRM$(\mathcal{U}^i)$
4 $\quad$ $q^i \leftarrow$ InitializeQ(M^i)
5 $\quad$ $\Lambda_\top^i \leftarrow \emptyset, \Lambda_\bot^i \leftarrow \emptyset$
6 **for** $n = 1$ **to** $NumEpisodes$ **do**
7 $\quad$ $t \leftarrow 0$
8 $\quad$ **for** $i = 1$ **to** N **do**
9 $\quad\quad$ $u_t^i \leftarrow u_0^i, s_t^i \leftarrow s_I^i, \lambda_t^i \leftarrow \emptyset, done^i \leftarrow \bot$
10 $\quad$ **while** $\exists j \in N$ such that $done^j = \bot$ **do**
11 $\quad\quad$ **for** $i = 1$ **to** N **do**
12 $\quad\quad\quad$ TrainAgent$(done^i, s_t^i, u_t^i, q^i, \lambda_t^i, \Lambda_G^i, \Lambda_I^i, M^i, \mathcal{U}^i)$
13 $\quad\quad$ $t \leftarrow t + 1$
14
15 **Procedure** TrainAgent$(done, s_t, u_t, q, \lambda_t, \Lambda_\top, \Lambda_\bot, M, \mathcal{U})$
16 $\quad$ **if** $done = \bot$ **then**
17 $\quad\quad$ $a \leftarrow$ GetAction(q_{u_t}, s_t)
18 $\quad\quad$ $s_{t+1}, isGoalAchieved \leftarrow$ EnvStep(s_t, a)
19 $\quad\quad$ $\mathcal{L}_{t+1} \leftarrow l(s_t, s_{t+1})$
20 $\quad\quad$ $\lambda_{t+1} \leftarrow \lambda_t \oplus \mathcal{L}_{t+1}$
21 $\quad\quad$ $r, u_{t+1} \leftarrow$ GetNextRMState$(M, u_t, \mathcal{L}_{t+1})$
22 $\quad\quad$ $q_{u_t} \leftarrow$ UpdateQ$(q_{u_t}, s_t, a, r, s_{t+1}, q_{u_{t+1}})$
23 $\quad\quad$ **for** $u \in \mathcal{U} \setminus \{u_t\}$ **do**
24 $\quad\quad\quad$ $r, u' \leftarrow$ GetNextRMState$(M, u, \mathcal{L}_{t+1})$
25 $\quad\quad\quad$ $q_u \leftarrow$ UpdateQ$(q_u, s_t, a, r, s_{t+1}, q_{u'})$
26
27 $\quad\quad$ **if** $isGoalAchieved$ **then**
28 $\quad\quad\quad$ $\Lambda_\top \leftarrow \Lambda_\top \cup \{\lambda_{t+1}\}$
29 $\quad\quad\quad$ $\Lambda_\bot \leftarrow \Lambda_\bot \cup$ GenerateIncompleteTraces(λ_{t+1})
30 $\quad\quad\quad$ $M_{new} \leftarrow$ LearnRM$(\mathcal{U}, \Lambda_\top, \Lambda_\bot)$
31 $\quad\quad\quad$ $done \leftarrow \top$
32 $\quad\quad$ **else if** $u_{t+1} = u_A$ **then**
33 $\quad\quad\quad$ $\Lambda_\bot \leftarrow \Lambda_\bot \cup \{\lambda_{t+1}\}$
34 $\quad\quad\quad$ $M_{new} \leftarrow$ LearnRM$(\mathcal{U}, \Lambda_\top, \Lambda_\bot)$
35 $\quad\quad\quad$ $done \leftarrow \top$
36 $\quad\quad$ **else if** $t + 1 = T$ **then**
37 $\quad\quad\quad$ $\Lambda_\bot \leftarrow \Lambda_\bot \cup \{\lambda_{t+1}\}$
38 $\quad\quad\quad$ $done \leftarrow \top$
39
40 $\quad\quad$ **if** $M \neq M_{new}$ **then**
41 $\quad\quad\quad$ $M \leftarrow M_{new}$
42 $\quad\quad\quad$ $q \leftarrow$ InitializeQ(M)
43
44 **Procedure** LearnRM$(\mathcal{U}, \Lambda_\top, \Lambda_\bot)$
45 $\quad$ $M \leftarrow$ RMLearner$(\mathcal{U}, \Lambda_\top, \Lambda_\bot)$
46 $\quad$ **while** $M = \bot$ **do**
47 $\quad\quad$ $\mathcal{U} \leftarrow \mathcal{U} \cup \{u_{|\mathcal{U}|-1}\}$
48 $\quad\quad$ $M \leftarrow$ RMLearner$(\mathcal{U}, \Lambda_\top, \Lambda_\bot)$
49 $\quad$ **return** M

---

2. If λ_{t+1} is an incomplete trace and u_{t+1} is the final state of the RM (1.32), then λ_{t+1} is a counterexample since reaching the final state of the RM should be associated with completing the task. In this case, we add λ_{t+1} to $\Lambda_\perp$.
3. If λ_{t+1} is an incomplete trace and the maximum number of steps per episode T has been reached (1.36), we add λ_{t+1} to $\Lambda_\perp$.

In the first two cases, a new RM is learnt once the example sets have been updated. The LearnRM routine (1.44-49) finds the RM with the fewest states that covers the example traces. The RM learner (here ILASP) is initially called using the current set of RM states $\mathcal{U}$ (1.45); then, if the RM learner finds no solution covering the provided set of examples, the number of states is increased by 1. This approach guarantees that the RMs learnt at the end of the process are *minimal* (i.e., consist of the fewest possible states) [13]. Finally, the Q-functions associated with an RM are reset when a new RM is learnt; importantly, the Q-functions depend on the RM structure and, hence, they are not easy to keep when the RM changes (1.42). In all three enumerated cases, the episode is interrupted.

Example 5. Given the sub-task assigned to the agent A_3 in the THREEBUTTONS environment described in Fig. 2c, we consider a valid goal trace $\lambda = \langle \{\}, \{G_B\},$ $\{A_3^{R_B}\}, \{A_3^{\neg R_B}\}, \{A_3^{R_B}\}, \{R_B\}\rangle$. The generated incomplete traces are:

$$\langle\{\}\rangle,$$
$$\langle\{\}, \{G_B\}\rangle,$$
$$\langle\{\}, \{G_B\}, \{A_3^{R_B}\}\rangle,$$
$$\langle\{\}, \{G_B\}, \{A_3^{R_B}\}, \{A_3^{\neg R_B}\}\rangle,$$
$$\langle\{\}, \{G_B\}, \{A_3^{R_B}\}, \{A_3^{\neg R_B}\}, \{A_3^{R_B}\}\rangle.$$

4   Experiments

We evaluate our approach by looking at two metrics as the training progresses: (1) the collective reward obtained by the team of agents and (2) the number of steps required for the team to achieve the goal. The first metric indicates whether the team learns how to solve the task by getting a reward of 1 upon completion. The second one estimates the quality of the solution, i.e. the lower the number of steps the more efficient the agents. Although the training is performed in isolation for each agent, the results presented here were evaluated with all the agents acting together in a shared environment. Additionally, we perform a qualitative assessment of the RM learnt by each agent.

For all the experiments, we show in *green* the performance for the case where RMs are known a priori and provided to the agents. This scenario, where we have perfect knowledge about the structure of each sub-task, is used as an upper bound for the results of our approach. In *pink*, we show the performances of our approach where the local RMs are learnt. The *yellow* curve shows the performance when the agents learn to perform their individual task without the RM construct.[1] The results presented show the mean value and the 95%-confidence interval of the metrics evaluated over 5 different random seeds in two grid-based environments of size 7×7. We ran the experiments on an EC2 C5.2XLARGE instance on *AWS* using *Python* 3.7.0. We use the state-of-the-art inductive logic programming system *ILASP* v4.2.0 [19] to learn RMs from traces with a $1h$ timeout.

4.1 THREEBUTTONS Task

We evaluate our approach in the THREEBUTTONS task described in Example 1. The number of steps required to complete the task and the collective reward obtained by the team of agents throughout training are shown in Fig. 3. We observe that regardless of whether the local RMs are handcrafted or learnt, the agents learn policies that, when combined, lead to the completion of the global task. Indeed, the obtained collective reward converges to the maximum achievable indicating that all the agents have learnt to execute their sub-tasks and coordinate to achieve the common goal. Learning the RMs also helps speed up the learning for each of the agents, significantly reducing the number of training iterations required to learn to achieve the global task. However, we note that the policy learnt without the RM framework requires fewer steps on average than when the RMs are used.

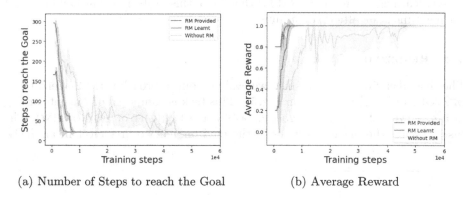

(a) Number of Steps to reach the Goal   (b) Average Reward

Fig. 3. Comparison between handcrafted RMs (RM Provided) and our approach learning the RMs from traces (RM Learnt) in the THREEBUTTONS environment.

We present in Fig. 4 the RM learnt by A_2 using our interleaved learning approach. We compare it to the "true" RM shown in Fig. 2b. Although important to truly reflect the dynamics of the environment and the labelling function, the back transition from u_3^2 to u_2^2 associated with the label $A_2^{\neg R_B}$ is missing

---

[1] This is equivalent of using a 2-state RM, reaching the final state only when the task is completed.

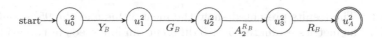

Fig. 4. Learnt RM for A_2.

from the learnt RM. In fact, A_2 must not only press the red button ($A_2^{R_B}$) but also remain on it until the event R_B is observed, or else R_B will never occur. The missing transition can be explained by the fact that the RM is learnt from *positive* (i.e., observable) examples only; that is, there are no *negative* (i.e., unfeasible) examples where the event R_B is observed directly after $A_2^{\neg R_B}$. This additional knowledge would have helped the RM learner identify this trace as impossible and realise that R_B can only be observed if A_2 does not stop pressing the red button ($A_2^{R_B}$). However, the joint learning of the RM and the associated optimal RL policies mitigates this problem. Due to the discounting factor and the fact that a non-zero reward is only given upon completing the task, the RL policies learn to minimize the number of steps required to achieve the task; hence, optimal policies will never consider travelling backwards in the RM. The RL policy associated with u_2^2 thus converges towards having A_2 staying on the red button until R_B is observed, removing the need to consider the back transition from u_3^2 to u_2^2.

Learning the global RM was also attempted in this environment. Unfortunately, learning the minimal RM for the size of this problem is hard and the RM learner timed out after the $1h$ limit.

4.2 RENDEZVOUS Task

The second environment in which we evaluate our approach is the 2-agent RENDEZVOUS task [21] presented in Fig. 5a. This task is composed of 2 agents acting in a grid with the ability to go UP, DOWN, LEFT, RIGHT or DO NOTHING. The task requires the agents to move in the environment to first meet in an *RDV*

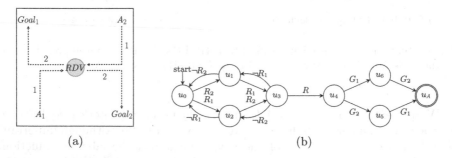

(a)   (b)

Fig. 5. Example of the RENDEZVOUS task where 2 agents must meet on the RDV point (green) before reaching their goal state $G1$ and $G2$ for agents $A1$ and $A2$ respectively. (Color figure online)

location, i.e. all agents need to simultaneously be in that location for at least one timestep. Then each of the agents must reach their individual goal location for the global task to be completed. The RM describing the task is shown in Fig. 5b.

We present in Fig. 6 the number of steps (6a) to complete the task and the collective reward received by the team of agents (6b) throughout training. The collective reward obtained while interleaving the learning of the policies and the learning of the local RMs converges to the maximum reward after a few iterations. In this task, the RM framework is shown to be crucial to helping the team solve the task in a timely manner. Indeed, without the RMs (either known a priori or learnt), the team of agents do not succeed at solving the task.

We also attempted to learn the global RM directly but the task could not be solved (i.e., a collective goal trace was not observed). The collective reward remained null throughout the training that we manually stopped after 1×10^6 timesteps.

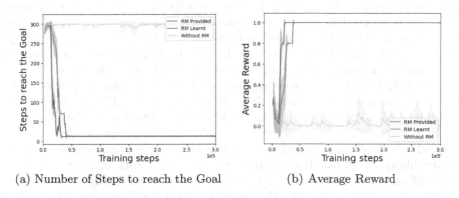

(a) Number of Steps to reach the Goal   (b) Average Reward

Fig. 6. Comparison between handcrafted RMs (RM Provided) and our approach learning the RMs from traces (RM Learnt) in the RENDEZVOUS environment.

5   Related Work

Since the recent introduction of RMs as a way to model non-Markovian rewards [27,28], several lines of research have based their work on this concept (or similar finite-state machines). Most of the work has focused on how to derive them from logic specifications [4] or demonstrations [5], as well as learning them using different methods [6,13–15,17,29,30].

In this work, based on the multi-agent with RMs work by Neary et al. [21], the RMs of the different agents are executed in parallel. Recent works have considered other ways of composing RMs, such as merging the state and reward transition functions of two RMs [9] or arranging them hierarchically by enabling RMs to call each other [14]. The latter is a natural way of extending this work since agents could share subroutines in the form of callable RMs.

The policy learning algorithm we employ for exploiting RMs is an extension of the QRM algorithm (see Sect. 2.2). However, algorithms based on making decisions at two hierarchical levels [13,27] or more [14] have been proposed. In the simplest case (i.e., two levels), the agent first decides which transition to satisfy from a given RM state, and then decides which low-level actions to take to satisfy that transition. For example, if agent A_2 gets to state u_3^2 in Fig. 2b, the decisions are: (1) whether to satisfy the transition labelled $A_2^{\neg R_B}$ or the one labelled R_B, and (2) given the chosen transition, act towards satisfying it. Unlike QRM, this method is not guaranteed to learn optimal policies; however, it enables lower-level sub-tasks to be reused while QRM does not. As the Q-functions learned by QRM depend on the structure of the whole RM, they must be re-initialized every time a new RM is learnt. While other methods attempt to leverage the previously learned Q-functions [30], the hierarchical approach outlined here does not need to re-learn all functions each time a new RM is induced.

In the multi-agent community, the use of finite-state machines for task decomposition in collaborative settings has been studied in [7,11] where a top-down approach is adopted to construct sub-machines from a global task. Unlike our work, these methods focus on the decomposition of the task but not on how to execute it. The algorithm presented in this paper interleaves both the decomposition and the learning of the policies associated with each state of the RM.

The MARL community studying the collaborative setting has proposed different approaches to decompose the task among all the different agents. For example, QMIX [22] tackles the credit assignment problem in MARL assuming that the team's Q-function can be factorized, and performs a value-iteration method to centrally train policies that can be executed in a decentralised fashion. While this approach has shown impressive empirical results, its interpretation is more difficult than understanding the structure of the task using RM.

The use of RMs in the MARL literature is starting to emerge. The work of Neary et al. [21] was the first to propose the use of RMs for task decomposition among multiple agents. As already mentioned in this paper, this work assumes that the structure of the task is known 'a priori', which is often untrue. With a different purpose, Dann et al. [8] propose to use RMs to help an agent predict the next actions the other agents in the system will perform. Instead of modelling the other agents' plan (or program), the authors argue that RMs provide a more permissive way to accommodate for variations in the behaviour of the other agent by providing a higher-level mechanism to specify the structure of the task. In this work as well, the RMs are pre-defined and provided to the agent.

Finally, the work of Eappen and Jagannathan [10] uses a finite-state machine called 'task monitor', which is built from temporal logic specifications that encode the reward signal of the agents in the system. A task monitor is akin to an RM with some subtle differences like the use of registers for memory.

6   Conclusion and Future Work

Dividing a problem into smaller parts that are easier to solve is a natural approach performed instinctively by humans. It is also particularly suited for the multi-agent setting where all the available resources can be utilized to solve the problem or when a complementary set of skills is required.

In this work, we extend the approach by Neary et al. [21], who leverage RMs for task decomposition in a multi-agent setting. We propose a novel approach where the RMs are learnt instead of manually engineered. The method presented in this paper interleaves the learning of the RM from the traces collected by the agent and the learning of the set of policies used to achieve the task's goal. Learning the RMs associated with each sub-task not only helps deal with non-Markovian rewards but also provides a more interpretable way to understand the structure of the task. We show experimentally in the THREEBUTTONS and the RENDEZVOUS environments that our approach converges to the maximum reward a team of agents can obtain by collaborating to achieve a goal. Finally, we validated that the learnt RM corresponds indeed to the RM needed to achieve each sub-task.

While our approach lifts the assumption about knowing the structure of the task 'a priori', a set of challenges remain and could be the object of further work:

1. A labelling function is assumed to exist for each of the agents and be known by them. In other words, each agent knows the labels relevant to its task and any labels used for synchronization with the other team members.
2. Noisy labelling functions are currently not supported by our approach but could be the object of future work. Extending the RM framework with probabilistic transition would be required adding another level of complexity to the RM learning process.
3. Our approach assumes that the global task has been automatically divided into several tasks, each to be completed by one of the agents. Dynamically creating these sub-tasks and assigning them to each of the agents is crucial to having more autonomous agents and would make an interesting topic for further research.
4. We leverage task decomposition to learn simpler RMs that when run in parallel represent a more complex global RM. Task decomposition can be driven further through task hierarchies, which could be represented through hierarchies of RMs [14]. Intuitively, each RM in the hierarchy should be simpler to learn and enable reusability and further parallelization among the different agents.
5. Finally, to allow even more reusability among the different agents and learn "skills" instead of tasks. The framework could be extended to learn transition using first-order formulae rather than propositions. This could potentially reduce the number of RL policies to learn and allow generalization of the skills learnt.

The use of RMs in the multi-agent context is still in its infancy and there are several exciting avenues for future work. We have demonstrated in this paper that it is possible to learn them under certain assumptions that could be relaxed in future work.

Acknowledgements. Research was sponsored by the Army Research Laboratory and was accomplished under Cooperative Agreement Number W911NF-22-2-0243. The views and conclusions contained in this document are those of the authors and should not be interpreted as representing the official policies, either expressed or implied, of the Army Research Laboratory or the U.S. Government. The U.S. Government is authorised to reproduce and distribute reprints for Government purposes notwithstanding any copyright notation herein.

References

1. Albrecht, S.V., Christianos, F., Schäfer, L.: Multi-Agent Reinforcement Learning: Foundations and Modern Approaches. MIT Press, Cambridge (2023)
2. Ardon, L., Vadori, N., Spooner, T., Xu, M., Vann, J., Ganesh, S.: Towards a fully RL-based market simulator. In: Proceedings of the ACM International Conference on AI in Finance (ICAIF), pp. 7:1–7:9 (2021)
3. Busoniu, L., Babuska, R., De Schutter, B.: A comprehensive survey of multiagent reinforcement learning. IEEE Trans. Syst. Man Cybern. Part C (Appl. Rev.) **38**(2), 156–172 (2008)
4. Camacho, A., Toro Icarte, R., Klassen, T.Q., Valenzano, R.A., McIlraith, S.A.: LTL and beyond: formal languages for reward function specification in reinforcement learning. In: Proceedings of the International Joint Conference on Artificial Intelligence (IJCAI), pp. 6065–6073 (2019)
5. Camacho, A., Varley, J., Zeng, A., Jain, D., Iscen, A., Kalashnikov, D.: Reward machines for vision-based robotic manipulation. In: Proceedings of the IEEE International Conference on Robotics and Automation (ICRA), pp. 14284–14290 (2021)
6. Christoffersen, P.J.K., Li, A.C., Toro Icarte, R., McIlraith, S.A.: Learning symbolic representations for reinforcement learning of non-Markovian behavior. In: Proceedings of the Knowledge Representation and Reasoning Meets Machine Learning (KR2ML) Workshop at the Advances in Neural Information Processing Systems (NeurIPS) Conference (2020)
7. Dai, J., Lin, H.: Automatic synthesis of cooperative multi-agent systems. In: Proceedings of the IEEE Conference on Decision and Control (CDC), pp. 6173–6178 (2014)
8. Dann, M., Yao, Y., Alechina, N., Logan, B., Thangarajah, J.: Multi-agent intention progression with reward machines. In: Proceedings of the International Joint Conference on Artificial Intelligence (IJCAI), pp. 215–222 (2022)
9. De Giacomo, G., Favorito, M., Iocchi, L., Patrizi, F., Ronca, A.: Temporal logic monitoring rewards via transducers. In: Proceedings of the International Conference on Principles of Knowledge Representation and Reasoning (KR), pp. 860–870 (2020)
10. Eappen, J., Jagannathan, S.: DistSPECTRL: distributing specifications in multi-agent reinforcement learning systems. arXiv preprint arXiv:2206.13754 (2022)
11. Elsefy, A.E.: A task decomposition using (HDec-POSMDPs) approach for multi-robot exploration and fire searching. Int. J. Robot. Mechatron. **7**(1), 22–30 (2020)
12. Fuchs, F., Song, Y., Kaufmann, E., Scaramuzza, D., Durr, P.: Super-human performance in gran turismo sport using deep reinforcement learning. IEEE Robot. Autom. Lett. **6**(3), 4257–4264 (2021)
13. Furelos-Blanco, D., Law, M., Jonsson, A., Broda, K., Russo, A.: Induction and exploitation of subgoal automata for reinforcement learning. J. Artif. Intell. Res. **70**, 1031–1116 (2021)

14. Furelos-Blanco, D., Law, M., Jonsson, A., Broda, K., Russo, A.: Hierarchies of reward machines. In: Proceedings of the International Conference on Machine Learning (ICML), pp. 10494–10541 (2023)
15. Gaon, M., Brafman, R.I.: Reinforcement learning with non-Markovian rewards. In: Proceedings of the AAAI Conference on Artificial Intelligence (AAAI), pp. 3980–3987 (2020)
16. Gold, E.M.: Complexity of automaton identification from given data. Inf. Control **37**(3), 302–320 (1978)
17. Hasanbeig, M., Jeppu, N.Y., Abate, A., Melham, T., Kroening, D.: DeepSynth: automata synthesis for automatic task segmentation in deep reinforcement learning. In: Proceedings of the AAAI Conference on Artificial Intelligence (AAAI), pp. 7647–7656 (2021)
18. Kaelbling, L.P.: Learning to achieve goals. In: Proceedings of the International Joint Conference on Artificial Intelligence (IJCAI), pp. 1094–1099 (1993)
19. Law, M., Russo, A., Broda, K.: The ILASP System for Learning Answer Set Programs (2015). www.ilasp.com
20. Mnih, V., et al.: Human-level control through deep reinforcement learning. Nature **518**(7540), 529–533 (2015)
21. Neary, C., Xu, Z., Wu, B., Topcu, U.: Reward machines for cooperative multi-agent reinforcement learning. In: Proceedings of the International Conference on Autonomous Agents and Multiagent Systems (AAMAS), pp. 934–942 (2021)
22. Rashid, T., Samvelyan, M., De Witt, C.S., Farquhar, G., Foerster, J., Whiteson, S.: Monotonic value function factorisation for deep multi-agent reinforcement learning. J. Mach. Learn. Res. **21**(1), 7234–7284 (2020)
23. Shalev-Shwartz, S., Shammah, S., Shashua, A.: Safe, multi-agent, reinforcement learning for autonomous driving. arXiv preprint arXiv:1610.03295 (2016)
24. Silver, D., et al.: Mastering the game of go with deep neural networks and tree search. Nature **529**(7587), 484–489 (2016)
25. Sultana, N.N., Meisheri, H., Baniwal, V., Nath, S., Ravindran, B., Khadilkar, H.: Reinforcement learning for multi-product multi-node inventory management in supply chains. arXiv preprint arXiv:2006.04037 (2020)
26. Sutton, R.S., Barto, A.G.: Reinforcement Learning: An Introduction. MIT Press, Cambridge (2018)
27. Toro Icarte, R., Klassen, T., Valenzano, R., McIlraith, S.: Using reward machines for high-level task specification and decomposition in reinforcement learning. In: Proceedings of the International Conference on Machine Learning (ICML), pp. 2107–2116 (2018)
28. Toro Icarte, R., Klassen, T.Q., Valenzano, R., McIlraith, S.A.: Reward machines: exploiting reward function structure in reinforcement learning. J. Artif. Intell. Res. **73**, 173–208 (2022)
29. Toro Icarte, R., Waldie, E., Klassen, T.Q., Valenzano, R.A., Castro, M.P., McIlraith, S.A.: Learning reward machines for partially observable reinforcement learning. In: Proceedings of the Advances in Neural Information Processing Systems (NeurIPS) Conference, pp. 15497–15508 (2019)
30. Xu, Z., et al.: Joint inference of reward machines and policies for reinforcement learning. In: Proceedings of the International Conference on Automated Planning and Scheduling (ICAPS), pp. 590–598 (2020)

Discrete-Choice Multi-agent Optimization: Decentralized Hard Constraint Satisfaction for Smart Cities

Srijoni Majumdar$^{(\boxtimes)}$ ⓘ, Chuhao Qin ⓘ, and Evangelos Pournaras ⓘ

School of Computing, University of Leeds, Leeds, UK
{s.majumdar,sccq,e.pournaras}@leeds.ac.uk

Abstract. Making Smart Cities more sustainable, resilient and democratic is emerging as an endeavor of satisfying hard constraints, for instance meeting net-zero targets. Decentralized multi-agent methods for socio-technical optimization of large-scale complex infrastructures such as energy and transport networks are scalable and more privacy-preserving by design. However, they mainly focus on satisfying soft constraints to remain cost-effective. This paper introduces a new model for decentralized hard constraint satisfaction in discrete-choice combinatorial optimization problems. The model solves the cold start problem of partial information for coordination during initialization that can violate hard constraints. It also preserves a low-cost satisfaction of hard constraints in subsequent coordinated choices during which soft constraints optimization is performed. Strikingly, experimental results in real-world Smart City application scenarios demonstrate the required behavioral shift to preserve optimality when hard constraints are satisfied. These findings are significant for policymakers, system operators, designers and architects to create the missing social capital of running cities in more viable trajectories.

Keywords: Decentralized Architectures · Hard Constraints · Global Cost function · Multi Agent Systems

1 Introduction

Setting hard constraints in how we consume, produce, distribute and manage urban resources becomes paramount for the sustainability of our cities [9]. Coordinated responses to climate change often aim to satisfy hard constraints for carbon footprint emissions and net-zero [28]. Smart Grid technologies are still under development because of challenges to satisfy hard operational constraints that can cause catastrophic power blackouts [23] (see Fig. 1). The satisfaction of hard constraints is also the safeguards for safety and the social capital for trust in establishing autonomous self-driving cards at scale [7]. Currently, the complexity, scale and decentralization of socio-technical infrastructures in Smart Cities are a barrier for satisfying hard constraints by design. Instead, soft constraints prevail in the vast majority of optimization and learning approaches

© The Author(s), under exclusive license to Springer Nature Switzerland AG 2024
F. Amigoni and A. Sinha (Eds.): AAMAS 2023 Workshops, LNAI 14456, pp. 60–76, 2024.
https://doi.org/10.1007/978-3-031-56255-6_4

applied to the broader spectrum of Smart City applications [13,17,21,24]. This research gap is the focus and subject of this paper.

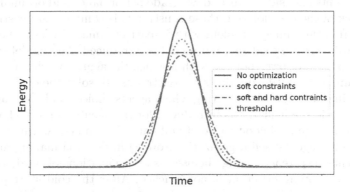

Fig. 1. An illustrative optimization scenario. *Baseline*: Without any optimized demand response, the power peak causes a blackout. *Soft constraints*: Significant power peak reduction that does not though prevent the power blackout. *Hard constraints*: Guarantee the reduction of the power peak below the blackout threshold.

This paper studies the decentralized hard constraint satisfaction in discrete-choice multi-agent optimization, in particular distributed multi-objective combinatorial optimization problems. This is a large class of problems [21] in which agents autonomously determine a number of finite options to choose from (operational flexibility). The agents may have their own preferences over these alternatives, expressed with a *discomfort cost* for each option. However, such choices often turn out to be inter-dependent to minimize system-wide *inefficiency and unfairness costs* (non-linear cost functions) for which the agents may have (explicitly or implicitly) interest as well. These choices require coordination and computing the optimal combination of choices in an NP-hard problem [24]. This is especially the case when agents come with different levels of *selfish vs. altruistic* behavior, with which they prioritize the minimization of their individual discomfort cost over the collective inefficiency and unfairness cost. Smart Cities are full of emerging application scenarios that can be modelled as such optimization problems [21,24,27]: power peak reduction to avoid blackouts, shift of power demand to consume more available renewable energy resources, coordinated vehicle routing to decrease travel times, traffic jams and air pollution, coordinated swarms of Unmanned Aerial Vehicles (UAVs) for distributed sensing, load-balancing of bike sharing stations, and other.

So far, heuristics for solving these distributed multi-agent optimization problems mainly address soft constraints, which is the best effort to minimize all involved costs. This is because in the absence of complete information in the agents, it is simpler to design algorithms that search efficiently for good solutions even if these are not the optimal ones. Instead, satisfying hard constraints opens up a Pandora box: without full information, any autonomous agent choice can violate the hard constraints that can only be satisfied with certainty at an aggregate level. Letting agents make conservative choices to avoid violating the hard constraints may significantly downgrade performance and optimality, while any rollback of choices violating these constraints is complex and costly.

To address this timely problem with impact on Smart Cities, a new decentralized hard constraint satisfaction model is introduced. The model constructs ranges of upper and lower bounds within which the aggregate choices and costs must remain, while optimizing for soft constraints. To solve the cold start problem in the initialization phase during which agents choices and costs are undergoing aggregation, a heuristic is introduced for the agents to make choices with the highest average likelihood to satisfy all hard constraints. As this heuristic is sensitive to the agents' self-organization (order) in decision-making and aggregation, agents keep reorganizing themselves after violations of hard constraints as long as a stopping criterion is not reached. After the cold start phase and after agents successfully satisfy the hard constraints, they can shift entirely to the optimization of the soft constraints while preserving the satisfaction of the hard constraints locally with a low cost.

The proposed model is integrated into a collective learning algorithm, the *Iterative Economic Planning and Optimized Selections* (I-EPOS) [24]. This allows a comprehensive assessment of the decentralized hard constraint satisfaction model and its impact on the optimality of the soft constraints. For the first time, experimental evaluation with real-world data from Smart City scenarios disentangle the performance sacrifice as a result of satisfying hard constraints and the additional agents' altruism level required to mitigate such sacrifice. These findings are highly revealing for system operators, policymakers, system designers and architects. They can inform them about the additional social capital (incentives/rewards) that they require to build (and pay) to preserve the cost-effectiveness of socio-technical infrastructures operating with hard constraints.

The contributions of this paper are summarized as follows: (i) A model of decentralized hard constraint satisfaction on optimizing aggregate agents' choices and their aggregate costs. (ii) The instantiation of this model on a decentralized multi-objective combinatorial optimization algorithm of collective learning for multi-agent systems. (iii) The applicability of decentralized hard constraint satisfaction on three Smart Cities scenarios using real-world data: energy, bike sharing and UAVs swarm sensing. (iv) Insights about the optimality sacrifice as moving from soft to hard constraints and how this optimality loss is measured in terms of the required behavioral shift to preserve performance, i.e. restoring altruism deficit. (v) An open-source software artifact implementation of the model for the I-EPOS collective learning algorithm.

This paper is summarized as follows: Sect. 2 reviews related methods. Section 3 introduces the decentralized hard constraint satisfaction model. Section 4 illustrates the applicability of this model to the collective learning algorithm of I-EPOS and its implementation. Section 5 illustrates the experimental methodology and the evaluation. Section 6 concludes this paper and outlines future work.

Table 1. Comparison of self adaptive decentralized approaches

Attributes	I-EPOS [24]	EPOS [20]	COHDA [10,17,18]	H-DPOP [4,14]
Plan Selection - Autonomy	Locally	Parent-Level	Locally	Parent-Level
Computational Cost	agent: $O(pt)$; system: $O(pt\ log\ a)$	agent: $O(p^c)$; system: $O(p^c\ log\ a)$	agent: $O(pt)$; system: $O(pt)$	agent: $O(p^c)$; system: $O(p^c)$
Communication Cost	agent: $O(t)$; system: $O(t\ log\ a)$	agent: $O(p)$; system: $O(p\ log\ a)$	agent: $O(at)$; system: $O(at)$	agent: $O(p^c)$; system: $O(p^c)$
Information Exchange	tree overlay; aggregate information;	tree overlay; aggregate messages;	k-connected graph; full information;	tree overlay; full information;
Soft Constraints	local (initialization), aggregated choices in global plan	local (initialization), aggregated choices in global plan	global	no soft constraints
Hard Constraints	local (initialization)	local (initialization)	local (initialization), global	local (initialization), global

p: number of plans (options), t: number of iterations, c: number of children, a: number of agents.

2 Comparison to Related Work

We study a discrete-choice distributed multi-objective optimization problem for multi-agent systems with both soft and hard constraints. In such systems, most related optimization approaches [12] operate as partially observable systems with agents communicating with their neighbors.

There are approaches that rely on an asynchronous hierarchical process using depth-first search to order agents that communicate with their parents to make choices and optimize objective functions [2]. Mailler et al. [15] cluster agents based on the constraints they attempt to satisfy, with a central controller that uses a branch and bound paradigm for searching solutions. These hierarchical approaches suffer from failure risks, performance bottlenecks, and potential privacy breaches in application scenarios involving sensitive personal data, e.g. location and health data.

Multi-agent reinforcement learning approaches with constraints on agent choices are earlier studied. For instance, Curran et al. [5] generate rewards for agents to optimize delay intervals that prevent air traffic congestion with greedy scheduling to implement *hard stop* (constraints) for agents when the delay surpasses a limit. Rollback to previous states (*warm restarts*) are earlier studied upon violation of global hard constraints in [19]. Simao et al. [29] learn non-violated execution by training using datasets containing constrained actions of the agents and corresponding global states of a centrally controlled environment. Even though the execution is decentralized, the learned model used to provide recommendations to the agents is an outcome of a centralized computation.

Decentralized and asynchronous versions of population search-based optimization methods, such as particle swarm optimization [1] or ant colony [8] algorithms show a slow convergence with high communication cost to rollback after violations of hard constraints, while improving global fitness and local search. This may slow down online real-time adaptations. Violation of hard constraints is prevented via message broadcasting that rolls back all choices made after a violation [2].

Other earlier approaches use tree overlay network structures for aggregation to aggregate messages from child nodes in form of hypercubes to reduce the frequency of message exchanges [4, 6, 14]. These methods use dynamic programming approach and thus storing all solutions increases the size of messages exponentially.

Other highly efficient discrete-choice multi-agent optimization methods, such as COHDA [10] and EPOS [24], address a large spectrum of NP-hard combinatorial problems in the domains of Smart Grids and Smart Cities [11, 17, 21, 25]. COHDA generalizes well in different communication structures among the agents that have full view of the systems, while EPOS focuses on hierarchical acyclic graphs such as trees to perform a cost-effective decision-making and aggregation of choices. Table 1 compares the design and efficiency of multi-agent optimization approaches, as well as how they address soft and hard constraints.

As COHDA shares full information between agents, it has higher communication cost. The computational cost is lower at global level for COHDA compared to EPOS because of the agents' brute force search to aggregate choices. Both COHDA and EPOS focus on satisfying soft constraints, like minimizing cost functions that satisfy balancing (minimum variance and standard deviation [24]) or matching (minimum root mean square error, residual sum of squares [25]) objectives. However, satisfying global hard constraints (Table 1) is challenging as agents need to additionally coordinate for choices, whose potential violations are only confirmed at an aggregate level, which makes any rollback of choices to avoid violations particularly complex. An expensive rollback procedure earlier introduced in COHDA [18] performs complete rescheduling using a 0–1 multiple-choice combinatorial to find another solution that satisfy the constraints.

Summarizing, satisfaction of hard constraints during initialization phase, when agents accumulate information about other agents' choices (cold start problem) remains an open challenge. It is also unclear how the satisfaction of hard constraints degrades the performance of these efficient algorithms based on their soft constraints. Addressing these open questions is the focus of this paper.

3   Hard Constraint Satisfaction Model

This section introduces the general optimization problem and the decentralized hard satisfaction model.

3.1   Optimization Problem

Table 2 summarizes the mathematical symbols of this article. Assume a socio-technical systems of n users, each assisted by a software agent, i.e. a personal digital assistant. Each agent i has a number of k options to choose from. Each option j is referred to as a *possible plan*, which is a sequence of real values $p_{i,j} = (p_{i,j,u})_{u=1}^{m} \in P_i = (p_{i,j})_{j=1}^{k}, \forall i \in \{1, ..., n\}$. Each agent selects one and only one plan $p_{i,s}$, which is referred to as the selected plan (i.e. the agent's choice). All agents' choices aggregate element-wise to the collective choice, the *global plan* $g = (g_u)_{u=1}^m = \sum_{i=1}^n p_{i,s}$ of the multi-agent system. A possible plan of an agent may represent a resource schedule or allocation, e.g. the energy consumed over time or the energy consumed from a certain supplier. Multiple possible plans for each agent represent alternatives and its operational flexibility. In the example of energy, the global plan represents the total energy consumption in the system over time or suppliers (see Fig. 1).

Table 2. Mathematical notations used in this paper.

Notation	Meaning
n	Number of agents
P_i	Set of possible plans for agent i
m	Plan size
k	Number of plans
$p_{i,j} \subset \mathbb{R}^m$	The jth plan as sequence of m elements of agent i
$p_{i,s}$	Selected plan of agent i
$g = \sum_{i=1}^n p_{i,s}$	Global plan from selected plans of all n agents
β_i	Discomfort factor for agent i
α_i	Unfairness factor for agent i
r	Constraints satisfaction rate
I	Inefficiency cost
D	Discomfort cost
U	Unfairness cost
$f_D : \mathbb{R}^m \to \mathbb{R}$	Discomfort cost function
$f_I : \mathbb{R}^m \to \mathbb{R}$	Inefficiency cost function
$f_U : \mathbb{R}^m \to \mathbb{R}$	Unfairness cost function
$\mathcal{U} \subset \mathbb{R}^m$	Sequence of upper bound hard constraints
$\mathcal{L} \subset \mathbb{R}^m$	Sequence of lower bound hard constraints
$\mathbb{E}(p_{i,j}, \mathcal{U})$	Expected satisfaction for upper bound constraints
$\mathbb{E}(p_{i,j}, \mathcal{L})$	Expected satisfaction for lower bound constraints

Agents' choices are made based on different, often opposing criteria. Each agent has its individual preferences over the possible plans, measured by the *discomfort cost* $f_D(p_{i,j}) = D_{i,j}$ of each plan j, which also makes the mean discomfort cost in the system $f_D(p_{1,s}, ..., p_{n,s}) = \frac{1}{n} \cdot \sum_{i=1}^n f_D(p_{i,s})$. Each agent can make independent choices to minimize their own discomfort cost. However, agents may also have interest to satisfy the following two general-purpose collective criteria: inefficiency cost $f_I(\sum_{i=1}^n p_{i,s}) = I_i$ and unfairness cost $f_U(D_{1,s}, ..., D_{n,s}) = U_i$. If these cost functions are non-linear, meaning the choices of the agents depend on each other, the satisfaction of soft constraints, i.e. minimizing the inefficiency and unfairness cost, is a combinatorial NP-hard optimization problem [24]. Balancing (e.g. min variance) and matching objectives (e.g. min residual of sum squares) are examples for measuring inefficiency cost with a broad applicability in load-balancing application scenarios of Smart Cities: minimizing power peaks, shifting demand to times with high availability of renewable energy resources, rerouting vehicles to avoid traffic jams, etc. Table 3 show such a case, in which three agents have two options (plans). These plans may represent energy consumption choices while forming an optimal global plans that meets the available energy supply. The elements may signify the power consumption for the day and night. The agents choose plans with minimum dispersion between elements (soft constraints), but that leads to a suboptimal global plan of [7, 13]. The global plan should also come with lower dispersion. The variance of the discomfort costs over the population of agents can measure the unfairness cost. Satisfying all of these (opposing) objectives depends on the selfish vs. altruistic behavior of the agents, e.g. whether they accept a plan with a bit higher discomfort cost to decrease inefficiency or unfairness cost. We can observe this in Table 3. If Agent C selects a plan ([6, 2]) with higher energy requirement during the day, it achieves to minimize the inefficiency cost and an optimum global plan of [10, 10] is achieved. Such multi-objective trade-offs are modelled with the parameters α_i and β_i such that:

$$p_{i,s} = \underset{j=1}{\overset{k}{argmin}}(1 - \alpha - \beta) \cdot f_\mathsf{I}(p_{1,s} + ... + p_{i,j} + ... + p_{n,s})$$
$$+ \alpha \cdot f_\mathsf{U}(D_{1,s}, ..., D_{i,j}, ..., D_{n,s})$$
$$+ \beta \cdot f_\mathsf{D}(D_{1,s}, ..., D_{i,j}, ..., D_{n,s}). \tag{1}$$

From the above equation, it is apparent that the choice of a plan cannot be easily optimized without (i) information of the other agents' choices and (ii) coordination of the agents' choices for non-linear cost functions that depend on each other. The optimization heuristics (Sect. 2) address the satisfaction of such soft constraints via various sequential information exchange, information aggregation and communication schemes that coordinate agents' choices. See the baseline scenario of soft constraints in Table 3.

However, introducing hard constraints on the aggregated choices g and their costs $D_{i,j}$, I_i, U_i is challenging. This is because there is no guarantee to satisfy the hard constraints in the absence of full information, which is usually the case for decentralized heuristics that require time to converge to full information. This is a particular cold start problem of initialization/exploration, during which the first choices are made under high uncertainty. As choices with high likelihood of violating hard constraints add up incrementally, it becomes increasingly hard to discover choices that will prevent such violations. Hence, agents need different and more conservative selection criteria that prioritize hard over soft constraints. Designing and evaluating these criteria is a contribution of this paper.

Table 3. An example of a discrete-choice combinatorial optimization problem with three agents ($n = 3$), each with two plans ($k = 2, m = 2$). All combinations of possible plan selections make $2^3 = 8$ possible global plans. Hard constraints with an upper bound on the aggregate choices (global plan g) are introduced with an expected satisfaction of $\sum_{u=1}^{m}(\mathcal{U}_u - p_{i,j,u})$. (1) The baseline scenario is the soft constraints that minimize the inefficiency cost $f_1(g) \approx |p_{i,j,1} - p_{i,j,2}|$. The global plan $[10, 10]$ is the one with the inefficiency cost. (2) Agents choose plans that maximize the expected satisfaction. This results in the global plan of $[6, 15]$ that satisfies the hard constraint $\mathcal{U} = [9,]$. (3) Similarly, the hard constraint $\mathcal{U} = [9,]$ is satisfied with the global plan of $[14, 9]$. (4) Both new hard constraints $\mathcal{U} = [10, 13]$ are satisfied with the global plan $[7, 13]$. (5) The second hard constraint $\mathcal{U} = [9, 9]$ is violated by the selected global plan $[7, 13]$. $\checkmark$: constraint satisfaction; $\times$: constraint violation in $p_{i,j}$ and g.

Constraints	Agent A		Agent B		Agent C		All Possible Global Responses							
Agents' Plans (p)	[3, 5]	[2, 7]	[1, 3]	[5, 2]	[6, 2]	[3, 5]	[10, 10]	[14, 9]	[7, 13]	[11, 12]	[9, 12]	[13, 11]	[6, 15]	[10, 14]
1. Soft Constraints (Baseline)														
Minimize Inefficiency Cost	$|3-5|=2$	$|2-7|=5$	$|1-3|=2$	$|5-2|=3$	$|6-2|=4$	$|3-5|=2$								
Selected Plans	$\checkmark$		$\checkmark$		$\checkmark$									
Selected Global Plan							$\checkmark$							
2. Hard Constraints - $\mathcal{U} = [9,]$														
Maximize Expected satisfaction	$(9-3)+0=6$	$(9-2)+0=7$	$(9-2)+0=7$	$(9-5)+0=4$	$(9-6)+0=3$	$(9-3)+0=6$								
Selected Plans		$\checkmark$	$\checkmark$		$\checkmark$									
Selected Global Plan													$\checkmark$	
3. Hard Constraints - $\mathcal{U} = [, 9]$														
Maximize expected satisfaction	$0+(9-5)=4$	$0+(9-7)=2$	$0+(9-3)=6$	$0+(9-2)=7$	$0+(9-2)=7$	$0+(9-5)=4$								
Selected Plans	$\checkmark$			$\checkmark$	$\checkmark$									
Selected Global Plan								$\checkmark$						
4. Hard Constraints - $\mathcal{U} = [10, 13]$														
Maximize expected satisfaction	$(10-3)+(13-5)=15$	$(10-2)+(13-7)=14$	$(10-1)+(13-3)=19$	$(10-5)+(13-2)=17$	$(10-6)+(13-2)=15$	$(10-3)+(13-5)=15$								
Selected Plans	$\checkmark$		$\checkmark$		$\checkmark$									
Selected Global Plan									$\checkmark$					
5. Hard Constraints - $\mathcal{U} = [9, 9]$														
Maximize expected satisfaction	$(9-3)+(9-5)=10$	$(9-2)+(9-7)=9$	$(9-1)+(9-3)=14$	$(9-5)+(9-2)=11$	$(9-6)+(9-2)=10$	$(9-3)+(9-5)=10$								
Selected Plans	$\checkmark$		$\checkmark$		$\checkmark$									
Selected Global Plan									$\times$					

3.2 A Heuristic for Satisfying Hard Constraints

The heuristic for satisfying the hard constraints on aggregate choices and their costs is illustrated in this section. Table 3 also illustrates an example of applying the heuristic in practice.

Constraints on Aggregate Choices. For each element g_u of a global plan g, a hard constraint is defined by a range (envelope) of an upper $\mathcal{U} = (\mathcal{U}_u)_{u=1}^{m}$ and lower $\mathcal{L} = (\mathcal{L}_u)_{u=1}^{m}$ bound, where $\mathcal{U}$, $\mathcal{L}$ are also sequences of real values of equal size $|\mathcal{U}| = |\mathcal{L}| = |g| = m$. Each value u of the upper bound denotes that $\mathcal{U}_u \geq g_u$, whereas for the lower bound it holds that $\mathcal{L}_u \leq g_u$.

The selected plan expected to satisfy all hard constraints at the initialization phase, during which the aggregate choices (global plan g) are not known, is estimated as follows:

$$p_{i,s} = \underset{p_{i,j} \in P_i}{argmax}\, \mathbb{E}(p_{i,j}, \mathcal{U}, \mathcal{L}), \tag{2}$$

where the expectation of satisfaction is given by:

$$\mathbb{E}(p_{i,j}, \mathcal{U}, \mathcal{L}) = \sum_{u=1}^{m}(\mathcal{U}_u - p_{i,j,u}) + \sum_{u=1}^{m}(p_{i,j,u} - \mathcal{L}_u). \tag{3}$$

Constraints on Aggregate Costs. The modeling for the hard constraints on the aggregate costs is exactly the same as the one of the aggregate choices, where the expected satisfaction for each of the costs of $f_D(p_{1,s}, ..., p_{n,s})$, $f_I(\sum_{i=1}^{n} p_{i,s})$ and $f_U(D_{1,s}, ..., D_{n,s})$ is calculated for upper and lower bounds with $|\mathcal{U}| = |\mathcal{L}| = 1$.

Constraint Satisfaction Rate. The effectiveness of the hard constraint satisfaction heuristic is measured by the satisfaction rate (r). This is the total number of satisfactions achieved out of a total number of trials made. These trials are often existing parameters of the optimization algorithms, for instance, random repetitions, or the order with which agents aggregate choices made to coordinate and optimize their own choices.

4   Hard Constraints Implementation

The model of decentralized hard constraint satisfaction is implemented and integrated into the I-EPOS collective learning algorithm[1] [24]. I-EPOS solves a large class of optimization problems, as formalized in Sect. 3.1. It is chosen due to its large spectrum of applicability in Smart City scenarios [21] as well as its superior performance in satisfying soft constraints [24]. Efficient coordinated choices are made using a self-organized [20] tree topology within which agents organize their interactions, information exchange and decision-making. I-EPOS benefits from the fact that trees are acyclic graphs: communication cost is very low and all exchanges in I-EPOS are at an aggregate level without double-counting. The coordination is a result of a more informed decision-making: each agent makes a choice taking into account the choices of a group of other agents (the tree branch underneath during initialization) or the choices of all agents at a previous time point (after initialization). Coordination evolves in multiple learning iterations, each consisting of a *bottom-up* and *top-down* phase. During the bottom-up phase, each agent chooses based on the new choices of the agents below and the choices of all agents in the previous iteration. However, each agent has an information gap: It has no information about the subsequent choices of the agents above in the tree. This problem is solved during the top-down phase, in which agents roll back (back propagation) to choices of the previous iteration as long as no costs reduction is achieved. Further information about the design of the I-EPOS collective learning algorithm is out of the scope of this paper and can be found in earlier work [24].

The decentralized hard constraint satisfaction model is implemented by filtering out the possible plans in Eq. 1 that violate the given upper and lower bounds. However, in the first learning iteration, it is not possible determine these plans that violate the hard constraints with certainty because each agent only knows about the aggregate choices of the agents underneath (and not the ones above). As a result, the root agent may end up having no possible plan that does not

---

[1] Available at: https://github.com/epournaras/EPOS/tree/hard_constraints.

violate the hard constraints. To prevent the likelihood of these violations, the agents make more conservative choices according to Eq. 2 during the first iteration, aiming at maximizing the expected satisfaction of the hard constraints. Once the hard satisfactions are satisfied, the agents switch back to plan selection according to Eq. 1, while keep filtering plans that violate the hard constraints. The agents cannot violate the hard constraints in these subsequent learning iterations because they always have the option to roll back to the choices made at the end of the first learning iteration during which hard constraints are satisfied (via the top-down phase).

Figure 2 illustrates the implementation of the hard constraints model on the open-source I-EPOS software artifact [16]. The implementation of the cost function interfaces is extended to filter out plans that violate the hard constraints, as well as the selection based on the expected satisfaction principle of Eq. 2. The hard constraints are controlled via the main input parameter file of I-EPOS (Java Properties). Constraints on aggregate choices and costs can be activated and deactivated. Two input .csv files are introduced, one for each type of hard constraints. Both contain the upper/lower bounds and the coding of the operators.

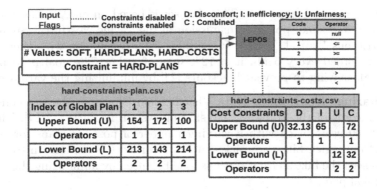

Fig. 2. Implementation of decentralized hard constraints satisfaction in I-EPOS [16].

5 Experimental Evaluation

Table 4 illustrates the application scenarios and settings of the experimental evaluation. A number of 1000 agents run the I-EPOS collective learning algorithm [24]. They are self-organized [20] in a height-balanced binary tree. The algorithm repeats 200 times, each time with a different random positioning of the agents in the tree. This introduces different decision-making order with which agents coordinate their optimized choices. At each repetition, the algorithm runs for 40 iterations, which is usually sufficient for convergence [24]. Evaluation is performed in three Smart City application scenarios: (i) energy demand-response, (ii) bike sharing and (iii) UAV swarm sensing. The optimized inefficiency cost function and the generation of plans are also outlined in Table 4.

Table 4. Parameters of the EPOS algorithm [24] for datasets

Parameter	Energy	Bike Sharing	UAV Swarm
Num. of agents (n)	1000	1000	1000
Num. of plans (k)	10	1 to 24	64
Plans size (m)	144	98	64
Num. of repetitions	200	200	200
Num. of iterations	40	40	40
Inefficiency cost f_1	Min VAR	Min VAR	Min RMSE

5.1 Smart City Application Scenarios

Energy Demand-Response. This dataset is based on energy disaggregation of the simulated zonal power transmission in the Pacific Northwest Smart Grid Demonstrations Project [22,25]. It contains 5600 consumers with their energy demand recorded every 5 min in a 12 h span of a day. The goal is to perform power peak shaving to prevent blackouts [24] by minimizing the variance.

Bike Sharing. The Hubway Data Visualization Challenge dataset consists of the trip records of the Hubway bike sharing system in Paris [22,24]. The data contain 2300 users, each with a varying number of possible plans for the bike stations from which bikes are picked up or returned (98 stations in total). The goal is to keep the bike sharing stations load balanced by minimizing the variance of the global plan.

UAV Swarm Sensing. The dataset contains 1000 drones that can capture images or videos of vehicle traffic information on public roadways over 64 areas of interest (sensing cells) that are uniformly distributed in the city map [22, 26]. Drones aim to collect the required amount of sensing data (target plan) determined by a continuous kernel density estimation, for instance, monitoring cycling risk based on past bike accident data [3].

5.2 Hard Constraint Satisfaction Works

For each application scenario, three incremental levels of hard constraints are set to the aggregate choices (global plan g). These levels are quantiles chosen empirically by observing the median global plan after several executions of I-EPOS based on soft constraints. The agents are assumed here altruistic, such that: $\beta_i = 0, \alpha_i = 0, \forall i \in \{1, ..., n\}$.

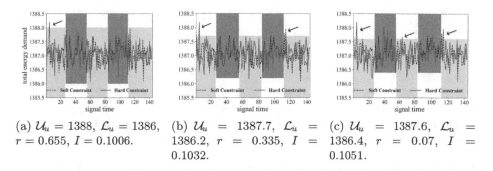

(a) $\mathcal{U}_u = 1388$, $\mathcal{L}_u = 1386$, $r = 0.655$, $I = 0.1006$.

(b) $\mathcal{U}_u = 1387.7$, $\mathcal{L}_u = 1386.2$, $r = 0.335$, $I = 0.1032$.

(c) $\mathcal{U}_u = 1387.6$, $\mathcal{L}_u = 1386.4$, $r = 0.07$, $I = 0.1051$.

Fig. 3. Optimization under soft and three levels of hard constraints in the energy demand-response scenario. Light-grey shaded areas represent the upper bound and dark-grey shaded areas the lower bound. Arrows point to violations of hard constraints.

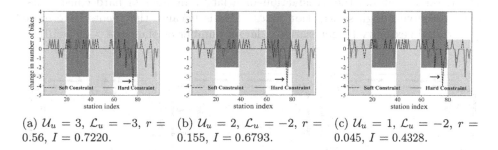

(a) $\mathcal{U}_u = 3$, $\mathcal{L}_u = -3$, $r = 0.56$, $I = 0.7220$.

(b) $\mathcal{U}_u = 2$, $\mathcal{L}_u = -2$, $r = 0.155$, $I = 0.6793$.

(c) $\mathcal{U}_u = 1$, $\mathcal{L}_u = -2$, $r = 0.045$, $I = 0.4328$.

Fig. 4. Optimization under soft and three levels of hard constraints in the bike sharing scenario. Light-grey shaded areas represent the upper bound and dark-grey shaded areas the lower bound. Arrows point to violations of hard constraints.

In the scenario of energy demand-response, hard constraints based on upper and lower bounds may represent an envelope of operation within which demand does not cause a blackout. In the scenario of bike sharing, hard constraints may represent limits on incoming or outgoing bikes that infrastructure operators may have, e.g. parking capacity. In the scenario of UAV swarm sensing, an upper bound of hard constraints may represent privacy-sensitive areas or regulated no-fly zones for drones. In contrast, a lower bound may represent minimal information required to monitor effectively a phenomenon, e.g. a forest fire or traffic jam.

Figures 3, 4 and 5 show the global plans for the soft constraint (baseline) along with three incremental and alternating levels of hard constraints (upper/lower bounds). Under soft constraints, the upper and lower bounds are violated, whereas hard constraints prevent these violations. Stricter hard constraints prevent more violations, however, the satisfaction rate (r) drops significantly, while the inefficiency cost increases. This shows that strict hard constraints are likely to oppose the soft constraints. Last but not least, note that the scenario of UAV

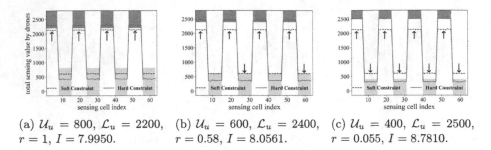

(a) $\mathcal{U}_u = 800$, $\mathcal{L}_u = 2200$,
$r = 1$, $I = 7.9950$.

(b) $\mathcal{U}_u = 600$, $\mathcal{L}_u = 2400$,
$r = 0.58$, $I = 8.0561$.

(c) $\mathcal{U}_u = 400$, $\mathcal{L}_u = 2500$,
$r = 0.055$, $I = 8.7810$.

Fig. 5. Optimization under soft and three levels of hard constraints in the UAV swarm sensing scenario. Light-grey shaded areas represent the upper bound and dark-grey shaded areas the lower bound. Arrows point to violations of hard constraints.

swarm sensing allows the satisfaction of a larger number of hard constraints, while preserving high satisfaction rates. This is because the agents have more operational flexibility by generating a larger number of plans ($64 > 24 > 10$).

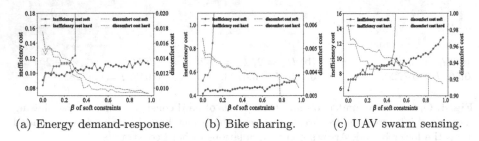

(a) Energy demand-response.   (b) Bike sharing.   (c) UAV swarm sensing.

Fig. 6. Inefficiency and discomfort cost as a function of the altruism level for soft and hard constraints for the three Smart City application scenarios.

5.3   Behavior Shift Can Mitigate Hard Constraints

Satisfying hard constraints results in a degrade of the performance profile (lower inefficiency cost) achieved under soft constraints. The recovery from this degrade is measured here as the required social capital (behavioral shift) that agents need to offer such that soft and hard constraints have equivalent performance. The raise of social capital is measured by the reduction of the mean β_i value in the population of agents that makes them more altruistic, see Eq. 1. The following method is introduced to measure the behavioral shift: I) Perform parameter sweep on I-EPOS under soft constraints for $\beta_i = 0$, to $\beta_i = 1, \forall i \in \{1, ...n\}$ with a step of 0.025. II) For each I-EPOS execution in Step 1 with a β_i value, a discomfort cost D and an inefficiency cost I, run I-EPOS under a hard constraint on the mean discomfort cost with an upper bound value equals to D (the

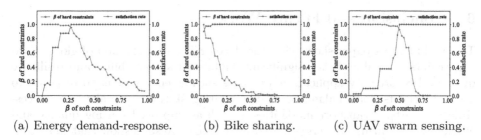

(a) Energy demand-response.   (b) Bike sharing.   (c) UAV swarm sensing.

Fig. 7. Required behavioral shift to mitigate the performance degrade of satisfying hard constraints for the three Smart City application scenarios. The satisfaction rate is also shown for each scenario. Performance comparison: β of soft constraints vs. β of hard constraints and satisfaction rate.

one of the I-EPOS execution under soft constraints). III) Derive the increased inefficiency cost under the hard constraint on the discomfort cost. IV) Find the β_i value from Step 1 that has the closest inefficiency cost with the one derived in Step 3 under the hard constraint and V) Compare the two β_i values in Step 2 and 4. The difference represents the required mean behavioral shift to mitigate the performance degrade of hard constraints.

Figure 6 illustrates the inefficiency and discomfort cost as a function of β_i under soft and hard constraints and the three different application scenarios. It becomes apparent that hard constraints require a minimum and significant level of altruism, otherwise, inefficiency cost rapidly explodes, especially in the scenarios without significant operational flexibility. This is also the reason why the discomfort cost becomes easier to reduce in the scenario of UAV swarm sensing, which comes with higher operational flexibility.

Figure 7 shows the required behavior shift to restore the performance loss as a result of satisfying hard constraints. For energy demand-response, the agents need on average 44.29% higher altruism under hard constraints to meet the performance of the soft constraints. The bike sharing scenario suggests an almost complete shift from selfish to altruistic behavior. Strikingly, the scenario of UAV swarm sensing shows performance gain as a result of satisfying hard constraints. As the number of plans is significantly higher for the UAV dataset, the search space is larger, which affects the optimality of the collective iterative learning paradigm in I-EPOS.

The satisfaction rate for the energy demand-response (Fig. 7a), bike sharing (Fig. 7b) and UAV swarm sensing (Fig. 7c) are 54.45%, 19.44% and 61.21% respectively. For $\beta \leq 0.475$ and $\beta \leq 0.25$, the satisfaction rate is 100% for the UAV swarm sensing and energy demand-response respectively. The operational flexibility by higher number of possible plans increases the constraints satisfaction rate.

6   Conclusion and Future Work

To conclude, this paper shows that the decentralized satisfaction of global hard constraints is feasible. It is a significant enabler for sustainability and resilience in several Smart City application scenarios such as energy demand-response to avoid blackouts, load balancing of bike sharing stations to make more accessible low-carbon transport modalities as well as improved sensing quality and efficiency by swarms of energy-constrained drones. Results show that hard constraints can be easily violated when optimizing exclusively for soft constraints. Instead, the proposed model prevents to a very high extent such violations.

Results also reveal the performance cost when hard constraints are introduced and how this cost can be mitigated by a behavioral shift towards a more altruistic behavior that sacrifices individual comfort for collective efficiency. These findings are invaluable for informing policy makers and systems operators of the required social capital that they need to raise to satisfy ambitious policies such as net-zero.

The open-source software artifact implementation of the proposed model to the I-EPOS collective learning algorithm is a milestone to encourage further research and application scenarios based on decentralized hard constraint satisfaction. Future work includes the applicability of the model to other decentralized optimization algorithms. The proposed heuristic is designed to satisfy all hard constraints together, which may be a limitation for high numbers of such opposing constraints, i.e. sacrifice of optimality and low satisfaction rates. Instead, a more incremental (and possibly partial) satisfaction of the hard constraints is part of future work. The further understanding of how to recover missing social capital to preserve both efficiency and fairness is also subject of future.

Acknowledgements. This work is supported by a UKRI Future Leaders Fellowship (MR/W009560/1): '*Digitally Assisted Collective Governance of Smart City Commons–ARTIO*', the Alan Turing Fellowship project '*New Edge-Cloud Infrastructure for Distributed Intelligent Computing*' and the SNF NRP77 'Digital Transformation' project "Digital Democracy: Innovations in Decision-making Processes", #407740_187249.

References

1. Akat, S.B., Gazi, V.: Decentralized asynchronous particle swarm optimization. In: 2008 IEEE Swarm Intelligence Symposium, pp. 1–8. IEEE (2008)
2. Billiau, G., Chang, C.F., Ghose, A.: SBDO: a new robust approach to dynamic distributed constraint optimisation. In: Desai, N., Liu, A., Winikoff, M. (eds.) PRIMA 2010. LNCS, vol. 7057, pp. 11–26. Springer, Cham (2012). https://doi.org/10.1007/978-3-642-25920-3_2
3. Castells-Graells, D., Salahub, C., Pournaras, E.: On cycling risk and discomfort: urban safety mapping and bike route recommendations. Computing **102**, 1259–1274 (2020)
4. Chen, Z., He, C., He, Z., Chen, M.: BD-ADOPT: a hybrid DCOP algorithm with best-first and depth-first search strategies. Artif. Intell. Rev. **50**, 161–199 (2018)

5. Curran, W.J., Agogino, A., Tumer, K.: Addressing hard constraints in the air traffic problem through partitioning and difference rewards. In: Proceedings of the 2013 International Conference on Autonomous Agents and Multi-Agent Systems, pp. 1281–1282 (2013)
6. Deng, Y., Chen, Z., Chen, D., Jiang, X., Li, Q.: PT-ISABB: a hybrid tree-based complete algorithm to solve asymmetric distributed constraint optimization problems. In: International Conference on Autonomous Agents and Multi-Agent Systems, pp. 1281–1282 (2019)
7. Du, H., Zhu, G., Zheng, J.: Why travelers trust and accept self-driving cars: an empirical study. Travel Behav. Soc. **22**, 1–9 (2021)
8. Gupta, A., Srivastava, S.: Comparative analysis of ant colony and particle swarm optimization algorithms for distance optimization. Procedia Comput. Sci. **173**, 245–253 (2020)
9. Helbing, D., et al.: Ethics of smart cities: towards value-sensitive design and co-evolving city life. Sustainability **13**(20), 11162 (2021)
10. Hinrichs, C., Lehnhoff, S., Sonnenschein, M.: COHDA: a combinatorial optimization heuristic for distributed agents. In: Filipe, J., Fred, A. (eds.) ICAART 2013. CCIS, vol. 449, pp. 23–39. Springer, Heidelberg (2014). https://doi.org/10.1007/978-3-662-44440-5_2
11. Hinrichs, C., et al.: A distributed combinatorial optimisation heuristic for the scheduling of energy resources represented by self-interested agents. Int. J. Bio-Inspired Comput. **10**(2), 69–78 (2017)
12. Kaddoum, E.: Optimization under constraints of distributed complex problems using cooperative self-organization. Ph.D. thesis (2011)
13. Khan, S., Paul, D., Momtahan, P., Aloqaily, M.: Artificial intelligence framework for smart city microgrids: state of the art, challenges, and opportunities. In: 2018 Third International Conference on Fog and Mobile Edge Computing (FMEC), pp. 283–288. IEEE (2018)
14. Kumar, A., Petcu, A., Faltings, B.: H-DPOP: using hard constraints for search space pruning in DCOP. In: AAAI, pp. 325–330 (2008)
15. Mailler, R., Lesser, V.: Solving distributed constraint optimization problems using cooperative mediation. In: Proceedings of the Third International Joint Conference on Autonomous Agents and Multiagent Systems. AAMAS 2004, pp. 438–445. IEEE (2004)
16. Majumdar, S., Qin, C., Pournaras, E.: Epos hard constraints support (2023). https://doi.org/10.5281/zenodo.7791326. www.zenodo.org/record/7791326
17. Nieße, A., Sonnenschein, M., Hinrichs, C., Bremer, J.: Local soft constraints in distributed energy scheduling. In: 2016 Federated Conference on Computer Science and Information Systems (FedCSIS), pp. 1517–1525. IEEE (2016)
18. Nieße, A., et al.: Conjoint dynamic aggregation and scheduling methods for dynamic virtual power plants. In: 2014 Federated Conference on Computer Science and Information Systems, pp. 1505–1514. IEEE (2014)
19. Parnika, P., Diddigi, R.B., Danda, S.K.R., Bhatnagar, S.: Attention actor-critic algorithm for multi-agent constrained co-operative reinforcement learning. In: International Conference on Autonomous Agents and MultiAgent Systems. ACM (2021)
20. Pournaras, E.: Multi-level reconfigurable self-organization in overlay services. Ph.D. thesis, Delft University of Technology. School of Technology Policy and Management (2013)

21. Pournaras, E.: Collective learning: a 10-year odyssey to human-centered distributed intelligence. In: 2020 IEEE International Conference on Autonomic Computing and Self-Organizing Systems (ACSOS), pp. 205–214. IEEE (2020)
22. Pournaras, E.: Agent-based planning portfolio (2023). www.figshare.com/articles/dataset/Agent-based_Planning_Portfolio/7806548
23. Pournaras, E., Espejo-Uribe, J.: Self-repairable smart grids via online coordination of smart transformers. IEEE Trans. Ind. Inf. **13**(4), 1783–1793 (2016)
24. Pournaras, E., Pilgerstorfer, P., Asikis, T.: Decentralized collective learning for self-managed sharing economies. ACM Trans. Auton. Adapt. Syst. (TAAS) **13**(2), 1–33 (2018)
25. Pournaras, E., Yao, M., Helbing, D.: Self-regulating supply-demand systems. Future Gener. Comput. Syst. **76**, 73–91 (2017)
26. Qin, C., Candan, F., Mihaylova, L., Pournaras, E.: 3, 2, 1, drones go! A testbed to take off UAV swarm intelligence for distributed sensing. In: Proceedings of the 2022 UK Workshop on Computational Intelligence. Springer (2022)
27. Qin, C., Pournaras, E.: Coordination of drones at scale: decentralized energy-aware swarm intelligence for spatio-temporal sensing. arXiv preprint arXiv:2212.14116 (2022)
28. Ramaswami, A., et al.: Carbon analytics for net-zero emissions sustainable cities. Nat. Sustain. **4**(6), 460–463 (2021)
29. Simão, T.D., et al.: AlwaysSafe: reinforcement learning without safety constraint violations during training. In: International Conference on Autonomous Agents and MultiAgent Systems. ACM (2021)

Deliberation and Voting
in Approval-Based Multi-winner Elections

Kanav Mehra[✉], Nanda Kishore Sreenivas[✉], and Kate Larson

University of Waterloo, Waterloo, ON, Canada
{kanav.mehra,nksreenivas,kate.larson}@uwaterloo.ca

Abstract. Citizen-focused democratic processes where participants deliberate on alternatives and then vote to make the final decision are quite popular today. While the computational social choice literature has extensively investigated voting rules, there is limited work that explicitly looks at the interplay of the deliberative process and voting. In this paper, we build a deliberation model using established models from the opinion-dynamics literature and study the effect of different deliberation mechanisms on voting outcomes achieved when using well-studied voting rules. Our results show that deliberation generally improves welfare and representation guarantees, but the results are sensitive to how the deliberation process is organized. We also show, experimentally, that simple voting rules, such as approval voting, perform as well as more sophisticated rules such as proportional approval voting [26] or equal shares [21] if deliberation is properly supported. This has ramifications on the practical use of such voting rules in citizen-focused democratic processes.

Keywords: Multi-winner Elections · Approval Voting · Deliberation

1 Introduction

Scenarios, where a committee must be selected to represent the interests of some larger group, are ubiquitous, ranging from political domains [6] to technical applications [25]. *Multi-winner* voting has been well studied with a focus on understanding how the 'best' committee can be selected. However, the properties desired in the selected committee would depend on the context and task requirements. The social choice literature has extensively investigated the quality of multi-winner voting rules with respect to notions of social welfare, representation, and proportionality [1,12,18,23,24]. We refer the reader to [13] for an extensive survey on the properties of multi-winner rules.

In citizen-focused democratic processes such as citizens' assemblies [10] and participatory budgeting [6], there exists extensive scope for discussion over the multitude of possible alternatives. For example, deliberation is an important phase in most implementations of participatory budgeting as it allows voters

---

Equal contribution.

F. Amigoni and A. Sinha (Eds.): AAMAS 2023 Workshops, LNAI 14456, pp. 77–93, 2024.
https://doi.org/10.1007/978-3-031-56255-6_5

to refine their preferences and facilitates the exchange of information, with the objective of reaching consensus [3]. Deliberation, specifically within social choice, has been individually studied through multiple approaches, ranging from theoretical studies introducing consensus-reaching deliberation protocols [9,11] to empirical research highlighting the positive effect of deliberation on voter preferences [20,22]. However, they do not investigate the impact of deliberation on the quantitative and qualitative properties of voting rules. While deliberation is a vital component of democratic processes [14,16], it cannot completely replace voting because, in reality, deliberation does not guarantee unanimity. A decision must still be made. Accordingly, we argue that it is essential to understand the relationship between voting and deliberation. To this end, we bridge the gap between deliberation and voting literature by experimentally studying the effect of deliberation on voting outcomes across different deliberation mechanisms.

In practice, participatory democratic processes must be simple and explainable to ensure citizen trust and engagement. Lack of transparency discourages participation, especially from under-represented communities. We argue that the "complexity" of a voting rule can be measured along three axes — computational complexity (for some voting rules it is computationally hard to determine the winning committee [2] while for others it is polynomial), the cognitive burden on the voter [4], and the ease of explaining the voting rule. Complicated rules may provide strong performance guarantees, but they are often hard to explain to the layperson. In this work, we argue that effective deliberation can circumvent the need for complicated voting rules and vastly improve voting outcomes even for simple rules such as classical approval voting (AV).

We focus on approval-based elections, where voters express preferences by sharing a subset of approved candidates. Approval ballots are used in practice due to their simplicity and flexibility [3–5]. They also offer scope for deliberation as voters are often left to decide between many alternatives. We present an agent-based model of deliberation and explore various alternatives for structuring deliberation groups. We evaluate standard multi-winner voting rules, both before and after voters have the opportunity to deliberate, with respect to standard objectives from the literature, including social welfare, representation, and proportionality. We show that deliberation, in almost all scenarios, significantly improves welfare, representation, and proportionality. However, the results are sensitive to the deliberation mechanism; increased exposure to diverse opinions (or agents from different backgrounds) enhances the quality of deliberation, achieves higher consensus, protects minority preferences, and in turn achieves better voting outcomes. Finally, our results indicate that in the presence of effective deliberation, *simple*, explainable voting rules such as approval voting perform as well as more sophisticated, *complex* rules. This can serve to guide the design and deployment of voting rules in citizen-focused democratic processes and support the development of democratic research platforms such as Ethelo, Polis, and LiquidFeedback[1]

---

[1] https://ethelo.com/, https://pol.is/home, https://liquidfeedback.com/en/.

2   Preliminaries

Let $E = (C, N)$ be an election, where $C = \{c_1, c_2, ..., c_m\}$ and $N = \{1, ..., n\}$ are sets of m candidates and n voters, respectively. Each voter $i \in N$, has an *approval ballot* $A_i \subseteq C$, containing the set of its approved candidates. The *approval profile* $A = \{A_1, A_2, ..., A_n\}$ represents the approval ballots for all voters. For a candidate $c_j \in C$, $N(c_j)$ is the set of voters that approve c_j and its *approval score*, $V(c_j) = |N(c_j)|$. Let $S_k(C)$ denote all k-sized subsets of the candidate set C. An *approval-based committee rule*, $R(A, k)$, is a social choice function that takes as input an approval profile A and committee size $k \in \mathbb{N}$ and returns a subset of candidates that form the winning committee $W_R \in S_k(C)$.

In this paper, we compare voting rules across three dimensions. **Utilitarian Social Welfare** objective measures the total overall 'utility' obtained from the elected committee. Formally, $SW(A, W) = \sum_{i \in N} \sum_{c \in W} u_i(c)$, where $u_i(c) \in \mathbb{R}$ is the utility voter i derives from candidate c. For a given rule, we compute its **utilitarian ratio** as $UR(R) = SW(A, W_R) / \max_{W \in S_k(C)} SW(A, W)$. **Representation Score** measures how many voters have at least one of their approved candidates in the final committee: $RP(A, W) = \sum_{i \in N} \min(1, |A_i \cap W|)$. We compute **representation ratio** as $RR(R) = RP(A, W_R) / \max_{W \in S_k(C)} RP(A, W)$. We also measure a **utility-representation aggregate score** $URagg(R) = UR(R) \cdot RR(R)$ to capture how well a voting rule balances both objectives. Finally, **proportionality** requires that a large enough voter group that collectively approves a shared candidate set must be "fairly represented". We use notions of extended and proportional justified representation (EJR and PJR, respectively) to check for proportionality (see Appendix A.1 for definitions). We count the number of instances that **satisfy EJR or PJR**.

We study the following approval-based **multi-winner voting rules**: Classical Approval Voting (AV), Approval Chamberlin-Courant (CC) [7], Proportional Approval Voting (PAV) [26], and Method-of-Equal-Shares (MES) [21]. They exhibit a wide range of properties, allowing for comparisons to be drawn across several axes. First, AV is known to maximize social welfare under certain conditions on voters' utility functions [18], however, there are no guarantees that AV provides proportionality [1]. Contrarily, CC maximizes representation, but its welfare properties are less well understood. Both PAV and MES guarantee EJR and maintain a balance between representation and social welfare. Finally, we argue that AV can be viewed as being *simple* in terms of computational complexity and explainability, whereas, PAV and MES are *complex* along at least one of these axes. Thus, this collection of rules covers the set of properties we are interested in. These voting rules are described in detail in Appendix A.2.

3   Deliberation

Our agent population N is divided into two sets — a *majority* and *minority*, where the number of agents in the majority is greater than that in the minority. Agents' initial preferences depend on their population group. Consistent with previous work [18], we assume an agent i's initial preference ranking, P_i^0, is sampled from a Mallows model [19], with reference rankings, Π_{maj} and Π_{min}, for the majority and minority populations, respectively. The rankings are then converted to an approval ballot using the top-ranked candidates. We further assume that agents have underlying cardinal utilities for candidates, consistent with their ordinal preferences.

The agents deliberate amongst themselves in an iterative process, where agents take turns being the speaker. The speaker makes its report (which reveals its thoughts and utilities for the candidates). All the other agents listen and update their utilities for all candidates based on a variation of the Bounded Confidence (BC) model [17]. We refer the reader to Appendix B for more details about the deliberation process.

In the real world, deliberation typically happens in small discussion groups [10]. To this end, we divide the population into g sub-groups of approximately equal size. Deliberation is conducted within these sub-groups where one *round* is complete when all agents in each group have spoken. The following strategies that we consider are informed by common heuristics used in practice.

Homogeneous Group: Each group contains only agents who are members of N_{maj} or N_{min}. That is, there is no mixing of minority and majority agents.

Heterogeneous Group: Each group is selected such that the ratio of the number of majority agents to the number of minority agents within the group is approximately equal to the majority:minority ratio in the overall population.

Random Group: Each group is created by randomly sampling agents from the population (without replacement) with equal probability.

Large Group: This is a special case where the deliberation process runs over the entire population of agents. It is infeasible in the real world, but we include this as a benchmark as it ensures maximum exposure to other agents' preferences.

Iterative Random: In each round, agents are randomly assigned to groups.

Iterative Golfer: This strategy is a variant of the social golfer problem. The number of rounds, R, is fixed *a priori*, and the number of times any pair of agents meet more than once is minimized. Please see Appendix D for details.

4   Experimental Evaluation

Our setup consists of 50 candidates ($|C| = 50$), 5 winners ($k = 5$), and 100 voters, with $N_{maj} = 80$ and $N_{min} = 20$. Agents' initial preferences are sampled using a Mallows model, with $\phi = 0.2$. To instantiate agents' utility functions, we generate m samples independently from the uniform distribution $\mathbf{U}(0,1)$, sort it, and then map the utilities to the candidates according to the agent's initial preference ranking P_i^0. When deliberating, agents are divided into 10 groups (except for the *large group* strategy). Iterative deliberation continues for $R = 5$ rounds. Please refer to Appendix E for more details on the setup.

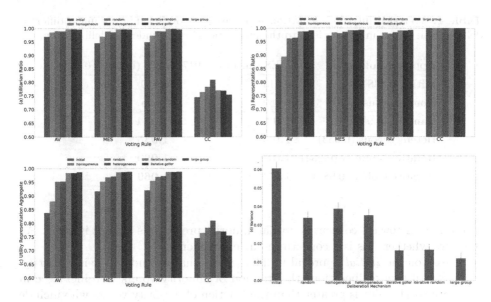

Fig. 1. Results for (a) Utilitarian ratio, (b) Representation Ratio, (c) Utility-Representation aggregate score, and (d) Average variance of agents' utilities for candidates (lower variance implies a higher degree of consensus in the population).

As a baseline, we apply every voting rule to the agent preferences *before* deliberation. We then run the different deliberation strategies and compute voting outcomes on the updated preferences. The average values over 10,000 simulations are reported. Figure 1 reports the impact of deliberation on voting outcomes. Due to space constraints, results for EJR and PJR satisfaction and other metrics have been moved to the appendix, Sect. F.

As expected, deliberation reduces disagreement amongst agents, moving all towards a consensus (Fig. 1(d)). Even a single round of deliberation improved outcomes across all voting rules and all objectives. However, the choice of the deliberation structure was important since *random* and *heterogeneous* consistently outperformed *homogeneous*. We hypothesize that this improvement was due to these deliberation strategies maximizing exposure to diverse opinions. By allowing majority and minority agents to interact, the minority agents had an opportunity to influence the majority population. This translated to better voting outcomes. In comparison to single-round methods, iterative deliberation further supports consensus and improves all objectives for most voting rules (except CC). CC's strong focus on coverage makes it unsuitable for deliberation methods that drive higher degrees of consensus since it fails to represent population groups proportionally (see Appendix G for a detailed discussion).

Minority Opinion Preservation: It is important to ensure that deliberation processes are inclusive and encourage minority participation [15]. While consensus would imply better voting outcomes, care must be taken to ensure that

Table 1. Average utility-representation aggregate score obtained by AV under different deliberation setups in comparison to the proportional rules under no deliberation.

Approval Voting	MES (initial) (0.917)	PAV (initial) (0.92)
Initial (0.838)	0.913	0.910
Homogeneous (0.88)	0.959	0.956
Random (0.952)	**1.038**	**1.034**
Heterogeneous (0.953)	**1.039**	**1.035**
Iterative Random (0.984)	**1.073**	**1.069**
Iterative Golfer (0.984)	**1.073**	**1.069**

when moving toward consensus, initial minority preferences are not ignored. We measure whether this is a concern in our experiments.

Based on the *initial* approval profile, we say that a candidate c is *minority-supported* if (pre-deliberation) the fraction of minority voters who include c in their approval ballot is greater than the fraction of majority voters who include c in their approval ballot. We then measure minority opinion preservation (MOP score) as the average number of pre-deliberation (initial) *minority-supported* candidates selected by AV (post-deliberation) across deliberation strategies. This serves as an indicator of whether minority preferences are *preserved*.

In the *initial* setup (no deliberation), AV does not elect any *minority-supported* candidates (i.e., MOP = 0). However, this improves as agents interact with the broader population. AV with single-round deliberation was better at preserving minority preferences (*homogeneous*, *random*, and *heterogeneous* achieve MOP of 0.2, 0.3, and 0.48, respectively). Iterative strategies exhibit further improvement with scores of 0.65 and 0.66 for *iterative random* and *golfer*, respectively. Finally, the *large group* setup achieves the highest MOP of 0.92. Thus, we show that AV with deliberation can *preserve* and represent minority preferences.

"Simple" vs. "Complex" Voting Rules: We compare AV with deliberation to MES and PAV without deliberation, using the utility-representation aggregate score ($URagg(R)$) as our measure (Table 1). Values greater than 1.0 indicate that AV with the corresponding deliberation mechanism achieves a better $URagg$ score than MES/PAV without deliberation. These findings support our argument that one does not necessarily have to use "complex" rules as "simple" rules coupled with effective deliberation strategies can be as effective.

5   Conclusion

We presented an empirical study of the relationship between deliberation and voting rules in approval-based multi-winner elections. Deliberation generally improves voting outcomes with respect to welfare, representation, and proportionality guarantees. Effectively designed mechanisms that increase exposure to diverse groups and opinions enhance the quality of deliberation, protect minority preferences, and in turn, achieve better outcomes. Importantly, we show that in the presence of effective deliberation, 'simpler' voting rules such as AV can be as powerful as more 'complex' rules without deliberation. We hope our findings can further support the design of effective citizen-focused democratic processes.

Appendix

A   Preliminaries and Definitions

A.1   Properties

We ideally want our voting rules to exhibit certain desired properties, representing the principles that should govern the selection of winners given individual ballots. In this paper, we compare voting rules across three dimensions: *social welfare, representation,* and *proportionality.* Intuitively, the *welfare* objective focuses on selecting candidates that garner maximum support from the voters. *Representation* cares about *diversity;* carefully selecting a committee that maximizes the number of voters represented in the winning committee.

It may not be possible to maximize both social welfare and representation, so *proportionality* serves as an important third objective to capture a compromise between welfare and representation. It requires that if a large enough voter group collectively approves a shared candidate set, then the group must be "fairly represented". Definitions of proportionality differ based on how they interpret "fairly represented".

Definition 1 (T-Cohesive Groups). *Consider an election $E = (C, N)$ with n voters and committee size k. For any integer $T \geq 1$, a group of voters N' is T-cohesive if it contains at least Tn/k voters and collectively approves at least T common candidates, i.e. if $|\cap_{i \in N'} A_i| \geq T$ and $|N'| \geq Tn/k$.*

Definition 2 (Proportional Justified Representation (PJR)). *A committee W of size k satisfies PJR if for each integer $T \in \{1, ..., k\}$ and every T-cohesive group $N' \subseteq N$, it holds that $|(\cup_{i \in N'} A_i) \cap W| \geq T$.*

Definition 3 (Extended Justified Representation (EJR)). *A committee W of size k satisfies EJR if for each integer $T \in \{1, ..., k\}$, every T-cohesive group $N' \subseteq N$ contains at least one voter that approves at least T candidates in W, i.e. for some $i \in N'$, $|A_i \cap W| \geq T$.*

A.2   Multi-winner Voting Rules

In this section, we define the set of approval-based multi-winner voting rules that form the basis of our analysis.

Approval Voting (AV): Given approval profile A and a committee W, the AV-score is $sc_{av}(A, W) = \sum_{c \in W} V(c)$. The AV rule is defined as $R_{AV}(A, k) =$

$\arg\max_{W \in S_k(C)} sc_{av}(A, W)$. This rule selects k candidates with the highest individual approval scores.

Approval Chamberlin-Courant (CC): The CC rule [7], $R_{CC}(A, k)$, picks committees that maximize representation score $RP(A, W)$. Given profile A, $R_{CC}(A, k) = \arg\max_{W \in S_k(C)} RP(A, W)$. It maximizes the number of voters with at least one approved candidate in the winning committee.

Proportional Approval Voting (PAV): [26] For profile A and committee W, the PAV-score is defined as $sc_{pav}(A, W) = \sum_{i \in N} h(|W \cap A_i|)$, where $h(t) = \sum_{i=1}^{t} 1/i$. The PAV rule is defined as $R_{PAV}(A, k) = \arg\max_{W \in S_k(C)} sc_{pav}(A, W)$. Based on the idea of diminishing returns, a voter's utility from having an approved candidate in the elected committee W decreases according to the harmonic function $h(t)$. It is a variation of the AV rule that ensures proportional representation, as it guarantees EJR [1]. PAV is the same as AV when committee size $k = 1$, but computing PAV is NP-hard [2].

Method-of-Equal-Shares (MES): $R_{MES}(A, k)$, also known in the literature as Rule-X [21], is an iterative process that uses the idea of budgets to guarantee proportionality. Each voter starts with a budget of k/n and each candidate is of unit cost. In round t, a candidate c is added to W if it is q-affordable, i.e. for some $q \geq 0, \sum_{i \in N(c)} \min(q, b_i(t)) \geq 1$, where $b_i(t)$ is the budget of voter i in round t. If a candidate is successfully added then the budget of each supporting voter is reduced accordingly. This process continues until either k candidates are added to the committee or it fails. If it fails, then another voting rule is used to select the remaining candidates.

B   Deliberation Models

In this section, we describe our agent population and the different deliberation processes we consider.

B.1   Voting Population: Preferences and Utilities

Our agent population N can be divided into two sets — a *majority* and *minority*, where the number of agents in the majority is greater than the number of agents in the minority. Agents' initial preferences depend on which group they belong to. In particular, we assume an agent i's initial preference ranking, P_i^0, is sampled from a Mallows model [19], with reference rankings, Π_{Maj} and Π_{Min}, for the majority and minority populations respectively.[2]

---

[2] The Mallows model is a standard noise model for preferences. It defines a probability distribution over rankings over alternatives (i.e. preferences), defined as $\mathbb{P}(r) = \frac{1}{Z} \phi^{d(r, \Pi)}$ where Π is a reference ranking, $d(r, \Pi)$ is the Kendall-tau distance between r and Π, and Z is a normalizing factor.

We assume that agents have underlying cardinal utilities for the candidates, denoted by a vector $U_i^t = <u_i^t(c_1), u_i^t(c_2), \ldots, u_i^t(c_m)>$, and we assume agents' utilities are bounded between 0 and 1.[3] U_i^0 is derived from the agent's initial preferences such that

$$\forall c_x, c_y \in C, u_i^0(c_x) \geq u_i^0(c_y) \text{ if } c_x \succ c_y \text{ in } P_i^0$$

The agents' utilities evolve over time as a function of the deliberation processes which are described next.

B.2   The Deliberation Process

Deliberation is defined as a "discussion in which individuals are amenable to scrutinizing and changing their preferences in the light of persuasion (but not manipulation, deception or coercion) from other participants" [8]. In this section, we describe the abstract deliberation process used by all agents. Consider a group of agents deliberating on the candidates. Each agent, i, announces its utilities, U_i^t, according to some randomly determined sequence.[4] After each announcement, every agent updates their own utilities, incorporating the information just received. We refer to the agent declaring its utilities at any given time as the *speaker*, and other agents in the group as *listeners*.

After the speaker has spoken, every listener incorporates the information shared (i.e. the speaker's utilities) and updates their own utilities. We use a variation of the Bounded Confidence (BC) model to capture these updates [17]. The Bounded Confidence model is a particularly good match for modelling deliberation in groups because it was intended to "describe formal meetings, where there is an effective interaction involving many people at the same time"[5]. In the BC model, listeners consider the speaker's report, and update their opinions (i.e. utilities for alternatives) only if the speaker's report is not "too far" from their own. The notion of distance is captured by a confidence parameter for each listener, Δ_i. Similar to recent extensions of the BC model[6], we use heterogeneous confidence levels, *i.e.*, different agents have different confidence levels. We refer the interested reader to the Appendix C for details about the original model.

The BC model was designed for one-dimensional opinion spaces. However, agents in our model discuss and update utilities derived from all m candidates in

---

[3] Some models assume unit utility if an elected candidate is on the approval ballot of the voter, and zero utility otherwise [18]. However, it is possible that a voter might derive some non-zero utility from an elected candidate even though it was not on the voter's ballot. Thus, we assume real-valued utilities between 0 and 1.

[4] As is common in much of the deliberation literature (e.g. [8,20]), we assume agents are non-strategic and truthfully reveal their utilities.

[5] Castellano, C., Fortunato, S., Loreto, V.: Statistical physics of social dynamics. Rev. Mod. Phys. 81, 591-646 (May 2009).

[6] Lorenz, J.: Continuous opinion dynamics under bounded confidence: A survey. International Journal of Modern Physics C 18(12), 1819-1838 (2007).

C, making it a multi-dimensional space. We make a simplifying assumption that agents' utilities for all m candidates are independent of each other, and apply the BC model to each dimension (candidate) independently.

We now describe the deliberation process in detail. Consider a group G^* and some arbitrary time t, when one of the agents in the group (denoted by x) is the speaker. After x has spoken, each listener ($i \in G^* - \{x\}$) updates its opinions for all candidates $c_j \in C$ using the following rule:

$$u_i^{t+1}(c_j) = \begin{cases} (1 - w_{ix})u_i^t(c_j) + w_{ix}u_x^t(c_j), & \text{if } |u_i^t(c_j) - u_x^t(c_j)| \leq \Delta_i \\ u_i^t(c_j), & \text{otherwise} \end{cases} \quad (1)$$

Recall that we are interested in heterogeneous agent populations, where there is a majority (N_{maj}) and minority (N_{min}) subset of agents, and agents within the same set have similar preferences. It is known that opinions from sources similar to oneself have a higher influence than opinions from dissimilar sources[7],[8]. To capture this phenomena, we introduce two different weights, α_i and β_i, $\alpha_i \geq \beta_i$, that are used in the update rule shown in Eq. 1. The choice of weight depends on the relationship between the speaker and the listener. If both speaker and listener belong to the same group, α_i is used, which means that the listener puts more weight on the speaker's utterance when updating its utility. If the speaker and listener belong to different groups, then β_i is used, meaning that the listener places less weight on the utterance of the speaker. In particular,

$$w_{ix} = \begin{cases} \alpha_i, & \text{if } \{i, x\} \subset N_{maj} \vee \{i, x\} \subset N_{min} \\ \beta_i, & \text{otherwise.} \end{cases} \quad (2)$$

C   Opinion Dynamics Models

We discuss two well-established models from opinion dynamics — DeGroot's classical model[9] and Hegselman and Krause's Bounded Confidence (BC) model [17].

According to DeGroot's classical model, an agent's updated opinion is simply the weighted sum of opinions from various sources (itself included). The weights were static, and could be different for different agents. So, for two agents x and y, x updates its opinion as:

$$x(t + 1) = w_{xx}x(t) + w_{xy}y(t) \quad (3)$$

[7] Mackie, D.M., Worth, L.T., Asuncion, A.G.: Processing of persuasive in-group messages. J Pers Soc Psychol 58(5), 812-822 (May 1990).

[8] Wilder, D.A.: Some determinants of the persuasive power of in-groups and out-groups: Organization of information and attribution of independence. Journal of Personality and Social Psychology 59(6), 1202-1213 (1990).

[9] Degroot, M.H.: Reaching a consensus. Journal of the American Statistical Association 69(345), 118-121 (1974).

where $x(t)$ denotes the opinion of agent x at time t, w_{xx} and w_{xy} denote x's weights on its own opinion and y's opinion, respectively. Note that the weights should sum up to 1, and therefore, $w_{xy} = 1 - w_{xx}$.

Later, there was the Bounded Confidence (BC) model [17] which introduced a global confidence level Δ. In the original paper, agents were on a network, and agents updated their opinions based on opinions of their neighbors. In the BC model, an agent x considered a neighbor's (y) opinion only if the neighbor's opinion was within x's confidence interval $[x(t) - \Delta, x(t) + \Delta]$. In the initial version, there were no distinct weights and all opinions within the confidence interval were weighted equally. When simplified for just two agents x and y, the opinion update for x is given by:

$$x(t+1) = \begin{cases} 1/2(x(t) + y(t)), & \text{if } y(t) \in [x(t) - \Delta, x(t) + \Delta] \\ x(t), & \text{otherwise} \end{cases} \qquad (4)$$

The BC model captures the idea of confirmation bias, and BC and its several modified versions have largely remained popular till date in the field of opinion dynamics.

D   Iterative Golfer

Iterative golfer strategy is a weaker version of the popular social golfer problem in combinatorial optimization[10],[11].

Social golfer problem: n golfers must be repeatedly assigned to g groups of size s. Find the maximum number of rounds (and the corresponding schedule) such that no two golfers play in the same group more than once.

Social golfer problem maximizes the number of rounds with a hard constraint that no two golfers should meet again. The iterative golfer strategy is a weaker version of this where we fix the number of rounds R, and minimize the number of occurrences where any pair of agents meet more than once. Given some group assignment $G^r = \{G_1^r, G_2^r, \ldots, G_g^r\}$ at round r, we introduce a cost given by:

$$cost(G^r) = \sum_{G_x \in G^r} \sum_{a,b \in G_x} f^2(a,b) \qquad (5)$$

where $f(a,b)$ is the number of times a and b have been in the same group in the previous rounds G^1 through G^{r-1}. The number of prior meetings is squared to ensure an even number of conflicts among all possible pairings (as opposed to one specific pair meeting repeatedly). We use an existing approximate solution[12]

[10] Liu, K., Löffler, S., Hofstedt, P.: Social golfer problem revisited. In: Agents and Artificial Intelligence (2019).

[11] Harvey, W.: CSPLib problem 010: Social golfers problem.

[12] https://github.com/islemaster/good-enough-golfers (Buchanan, B.: Good-enough golfers.).

that creates group assignment for each round such that the cost given by (5) is minimized. The iterative golfer can thus be seen as a more efficient strategy than iterative random if the objective is to ensure each agent has the highest possible exposure to others' preferences.

E   Further Details About the Experimental Setup

Our election setup consists of 50 candidates ($|C| = 50$)[13] and 100 voters, with 80 agents in the majority group (N_{maj}) and 20 in the minority group (N_{min}). Agents' initial preferences are sampled using a Mallows model, with $\phi = 0.2$. The reference ranking used while sampling a preference ordering depends on whether the agent belongs to N_{maj} or N_{min}. Reference rankings, Π_{maj} and Π_{min}, are sampled uniformly from all linear orders over C. Due to this sampling process, agents in either the majority or minority group have fairly similar preferences (as ϕ is relatively small) but the two groups themselves are distinct. To instantiate agents' utility functions, we generate m samples independently from the uniform distribution $\mathbf{U}(0, 1)$, sort it, and then map the utilities to the candidates according to the agent's preference ranking. We use a flexible ballot size b_i, where b_i is sampled from $\mathcal{N}(2k, 1.0)$. Agent i's approval ballot is the set consisting of top-b_i candidates from its preference ranking P_i. BC model parameters $(\Delta_i, \alpha_i, \beta_i)$ are sampled from uniform distributions over the full range for each parameter. We also ran experiments where all parameters were drawn from a normal distribution. There were no significant differences from the results reported here.

For our experiments, we use the Python library (*abcvoting*)[14] and use random tie-breaking when a voting rule returns multiple winning committees. To avoid trivial profiles, *i.e.*, profiles where an almost perfect compromise between welfare and representation is easily achievable, we impose some eligibility conditions. An initial approval profile A^0 is eligible only if $RR(AV, A^0) < 0.9 \wedge UR(CC, A^0) < 0.9$. This is a common technique used in simulations comparing voting rules based on synthetic datasets [18].

This entire simulation is repeated $10,000$ times and the average values are reported. To determine statistical significance while comparing any two sets of results, we used both the t-test and Wilcoxon signed-rank test, and we found the p-values to be roughly similar. All pairs of comparisons between deliberation group strategies for a given voting rule are statistically significant ($p < 0.05$) unless otherwise noted.

---

[13] Typically, project proposals are invited from the participants in PB [3,6]. So, there are a large number of candidate projects to choose from (e.g., PB instances in Warsaw, Poland had between 20-100 projects (36 on average). [12]).

[14] Lackner et al. abcvoting: A Python library of approval-based committee voting rules, 2021. Current version: https://github.com/martinlackner/abcvoting..

Table 2. EJR- and PJR-satisfaction (AV and CC).

Deliberation Strategy	EJR%		PJR%	
	AV	CC	AV	CC
Initial (no deliberation)	99.5	62.5	99.5	73.4
Homogeneous	96.4	69.9	96.4	75.1
Random	100	81.9	100	85.6
Heterogeneous	100	92.7	100	94.0
Iterative Random	100	31.4	100	53.6
Iterative Golfer	100	29.9	100	51.2
Large Group	100	6.10	100	23.4

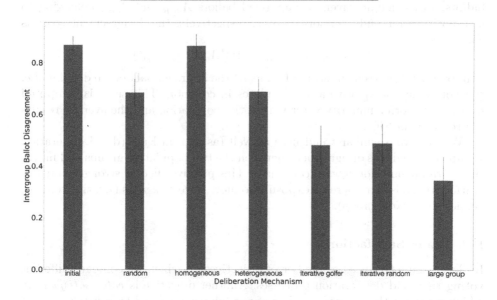

Fig. 2. Inter-group Ballot Disagreement

F   Further Results from Sect. 4

F.1   EJR and PJR Satisfaction

Table 2 shows the percentage of EJR- and PJR-satisfying committees returned by AV and CC. We focus only on AV and CC since the proportional rules MES and PAV guarantee EJR. Even under no deliberation (*initial*), AV satisfies EJR in almost all profiles, which further improves to perfect satisfaction with deliberation (except *homogeneous*). This is interesting since AV is not guaranteed to satisfy EJR.[15] EJR and PJR satisfaction for CC also improves if single-round deliberation is supported, with *heterogeneous* achieving the best result. Iterative deliberation, however, does not perform well. We believe that this arises due to CC's strong focus on representation (see Appendix G).

F.2   Inter-group Ballot Disagreement

In Fig. 1(d) we introduce a measure of consensus in the population as the average variance in agents' utilities and show that deliberation reduces disagreement amongst agents. To complement this analysis and further understand the impact of deliberation on agents' preferences, we introduce another metric that computes the disagreement between the majority and minority voters based on their ballots. In particular, given two approval ballots A_{min} and A_{maj} belonging to a minority voter and a majority voter, respectively, the disagreement score is computed as:

$$1 - (|A_{min} \cap A_{maj}| / \min(|A_{min}|, |A_{maj}|))$$

A maximum disagreement score of 1 means the approval ballots are disjoint, *i.e.* the voters do not approve any candidates in common. This score is computed for every majority-minority voter pair in the population and the average results are reported in Fig. 2.

We observe a similar trend here as well (as seen in Fig. 1(d)). Deliberation significantly reduces disagreement between the two population groups and moves the overall population toward consensus. This positive effect is stronger in deliberation methods that increase exposure to more, diverse agents (*i.e.* the iterative versions and *large group*).

F.3   Voter Satisfaction

In Sect. 2 we introduced a number of objectives on which we compare different voting rules and deliberation processes. Another objective is *voter satisfaction*, which measures the average number of candidates approved by a voter.
Voter Satisfaction Score: Given $W_R = R(A, k)$, the voter satisfaction is measured as the average number of candidates approved by a voter in W:

$$VS(R) = \frac{\sum_{i \in N} |A_i \cap W_R|}{|N|}. \qquad (6)$$

Figure 3 shows the average voter satisfaction obtained by the voting rules across different deliberation setups.

AV is expected to achieve the highest satisfaction since it picks candidates with the highest support, i.e. the average number of candidates approved by a

---

[15] Since the minority and majority agents have highly correlated approval sets, T-cohesive groups may exist only for a small set of minority- and majority-supported candidates, thereby making the EJR requirement easy to satisfy. Furthermore, previous research [12] shows that under many natural preference distributions (generated elections), there are many EJR-satisfying committees.

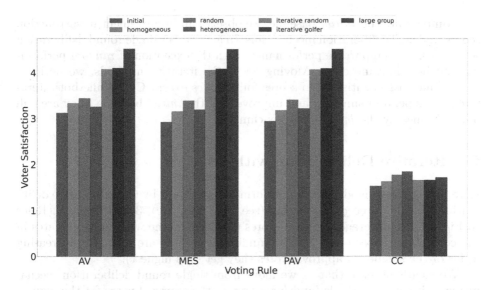

Fig. 3. Voter satisfaction achieved by the voting rules across deliberation mechanisms

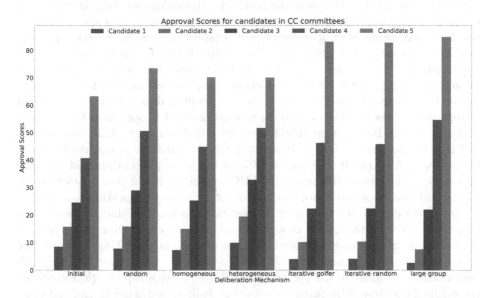

Fig. 4. Average approval scores obtained by the 5 candidates in the winning committee chosen by CC across different deliberation mechanisms.

voter will be high. MES and PAV achieve comparable scores, just slightly lower than AV. Finally, CC achieves the lowest satisfaction of all rules. In an attempt to maximize voter coverage, CC might choose winning candidates that represent few voters, and as a result, have low approval scores. Due to this, it maximizes diversity but achieves low voter satisfaction.

Compared to the *initial* baseline, we observe an improvement in satisfaction scores under all deliberation mechanisms. In general, all single round deliberation setups achieve comparable performance, with the exception of *random* performing the best in some cases. Moving on to the iterative methods, we notice a further increase in satisfaction scores for all rules except CC. While both iterative setups perform similarly and improve over the *initial* baseline, they are still outperformed by the *large group* benchmark.

G　Iterative Deliberation with CC

Here, we explain the odd drop in performance observed by CC in iterative deliberation and the *large group* setting (see Figs. 1(a), 1(c), 3 and Table 2). Refer to Fig. 4 for the average approval scores obtained by the winning candidates in the committees chosen by CC. The candidates (1 to 5) are ranked in increasing order of the number of approval votes they get (5 is highest).

We clearly observe that as we move from single round deliberation mechanisms to iterative methods (and *large group*), the approval votes for the highest supported candidate (5) increase and the same for the lowest supported candidate (1) decrease. For the iterative methods, approximately 80% of the agent population approves candidate_5 ($\approx$ 90% for *large group*). This also reinforces the fact that iterative deliberation approaches consensus, as a major proportion of voters approve a single candidate. Accordingly, CC is able to represent approximately 80% of the voters with just one candidate. Since CC only cares about maximizing voter coverage, it chooses the rest of the candidates to represent the remaining voters. This leads to sub-optimal outcomes since instead of representing the population groups proportionally, CC optimizes for coverage and chooses candidates that might have very little support. This can be seen in Fig. 4 as candidate_1 for the iterative methods and *large group* has less than 5% support. As a result, the almost 80% of the voter population that possibly gets only one representative in the final CC committee might be a cohesive voter group and thus, deserves more candidates for a fair and proportional outcome.

In conclusion, we see that with deliberation mechanisms that move towards consensus, CC exhibits a drop in welfare and proportionality guarantees since it is focused on maximizing representation. In general, other voting rules provide better overall performance than CC. However, if CC should ever be used with deliberation, we must pick an appropriate deliberation setup (single round) for the optimal outcome. This further shows that deliberation is not trivial and must be structured appropriately to obtain the best results.

References

1. Aziz, H., Brill, M., Conitzer, V., Elkind, E., Freeman, R., Walsh, T.: Justified representation in approval-based committee voting. Soc. Choice Welfare **48**(2), 461–485 (2017)
2. Aziz, H., Gaspers, S., Gudmundsson, J., Mackenzie, S., Mattei, N., Walsh, T.: Computational aspects of multi-winner approval voting. In: Workshops AAAI (2014)

3. Aziz, H., Shah, N.: Participatory budgeting: models and approaches. In: Pathways Between Social Science and Computational Social Science, pp. 215–236 (2021)
4. Benadè, G., Nath, S., Procaccia, A.D., Shah, N.: Preference elicitation for participatory budgeting. Manage. Sci. **67**(5), 2813–2827 (2021)
5. Brams, S., Fishburn, P.: Approval voting. Springer Science (2007)
6. Cabannes, Y.: Participatory budgeting: a significant contribution to participatory democracy. Environ. Urban. **16**(1), 27–46 (2004)
7. Chamberlin, J.R., Courant, P.N.: Representative deliberations and representative decisions: Proportional representation and the borda rule. American Political Science Review **77**(3), 718–733 (1983)
8. Dryzek, J.S., List, C.: Social choice theory and deliberative democracy: a reconciliation. British J. Political Sci. **33**(1), 1–28 (2003)
9. Elkind, E., Grossi, D., Shapiro, E., Talmon, N.: United for change: deliberative coalition formation to change the status quo. In: AAAI, pp. 5339–5346 (2021)
10. Elstub, S., Escobar, O., Henderson, A., Thorne, T., Bland, N., Bowes, E.: Citizens' assembly of Scotland (2022)
11. Fain, B., Goel, A., Munagala, K., Sakshuwong, S.: Sequential deliberation for social choice. In: International Conference on Web and Internet Economics (2017)
12. Fairstein, R., Vilenchik, D., Meir, R., Gal, K.: Welfare vs. representation in participatory budgeting. In: AAMAS, pp. 409–417 (2022)
13. Faliszewski, P., Skowron, P., Slinko, A., Talmon, N.: Multiwinner voting: a new challenge for social choice theory. Trends in Computational Social Choice 74 (2017)
14. Fishkin, J.: When the people speak: Deliberative democracy and public consultation. Oxford University Press (2009)
15. Gherghina, S., Mokre, M., Miscoiu, S.: Introduction: democratic deliberation and under-represented groups. Political Stud. Rev. **19**(2), 159–163 (2021)
16. Habermas, J.: Between Facts and Norms: Contributions to a Discourse Theory of Law and Democracy. Polity (1996)
17. Hegselmann, R., Krause, U.: Opinion dynamics and bounded confidence models. Anal. Simulation. JASSS **5**(3), 1–2 (2002)
18. Lackner, M., Skowron, P.: Utilitarian welfare and representation guarantees of approval-based multiwinner rules. Artif. Intell. **288**, 103366 (2020)
19. Mallows, C.L.: Non-null Ranking Models. I. Biometrika **44**(1–2), 114–130 (1957)
20. Perote-Peña, J., Piggins, A.: A model of deliberative and aggregative democracy. Econ. Philosophy **31**(1), 93–121 (2015)
21. Peters, D., Skowron, P.: Proportionality and the limits of welfarism. EC (2020)
22. Rad, S.R., Roy, O.: Deliberation, single-peakedness, and coherent aggregation. American Political Sci. Rev. **115**(2), 629–648 (2021)
23. Sánchez-Fernández, L., et al.: Proportional justified representation. In: AAAI (2017)
24. Skowron, P.: Proportionality degree of multiwinner rules. In: ACM-EC (2021)
25. Skowron, P., Lackner, M., Brill, M., Peters, D., Elkind, E.: Proportional rankings. In: IJCAI-17 (2017)
26. Thiele, T.N.: Om flerfoldsvalg. Oversigt over det Kongelige Danske Videnskabernes Selskabs Forhandlinger **1895**, 415–441 (1895)

Visionary Papers

Plan Generation via Behavior Trees Obtained from Goal-Oriented LTLf Formulas

Aadesh Neupane$^{(\boxtimes)}$ ⓘ, Michael A. Goodrich ⓘ, and Eric G. Mercer ⓘ

Brigham Young University, Provo, UT 84602, USA
adeshnpn@byu.edu, {mike,egm}@cs.byu.edu
https://cs.byu.edu

Abstract. Temporal logic can be used to formally specify autonomous agent goals, but synthesizing planners that guarantee goal satisfaction can be computationally prohibitive. This paper shows how to turn goals specified using a subset of *finite trace Linear Temporal Logic* (LTL_f) into a *behavior tree* (BT) that guarantees that successful traces satisfy the LTL_f goal. Useful LTL_f formulas for *achievement goals* can be derived using achievement-oriented task mission grammars, leading to missions made up of tasks combined using LTL operators. Constructing BTs from LTL_f formulas leads to a relaxed behavior synthesis problem in which a wide range of planners can implement the *action* nodes in the BT. Importantly, any successful trace induced by the planners satisfies the corresponding LTL_f formula. The usefulness of the approach is demonstrated in two ways: a) exploring the alignment between two planners and LTL_f goals, and b) solving a sequential *key-door* problem for a *Fetch* robot.

Keywords: Behavior Tree · Finite Linear Temporal Logic

1 Introduction

Specifying robot goals, creating plans, and verifying plans have received considerable attention in the research literature. Linear Temporal Logic (LTL) has been used in this context to specify system properties such as safety, liveness (something will keep happening), and goal-satisfaction [29]. However, the automata-based planning algorithms that accompany these systems often increase exponentially with an increase in the specification's length [24]. Moreover, the tight coupling between LTL verification and planning problems with automata-based controllers leads to expensive re-computation even for small specification changes.

One way to address these shortcomings is to decompose a complex goal specification into smaller modular specifications. Decomposing a complex goal specification using a Behavior Trees (BTs) is often useful because BTs are modular, maintainable, and reusable [11]. Prior work demonstrated polynomial time correct-by-construction BTs from an LTL formula [9], but a particularly restrictive specification format must be followed. The planning is also tightly coupled

F. Amigoni and A. Sinha (Eds.): AAMAS 2023 Workshops, LNAI 14456, pp. 97–122, 2024.
https://doi.org/10.1007/978-3-031-56255-6_6

with the BT decomposition algorithm, making the integration of off-the-shelf planners impossible.

This paper presents a mission-oriented grammar that generates LTL_f formulas appropriate for sequential *achievement goals* [34]. An achievement goal is defined as one that succeeds if a certain postcondition is satisfied without violating constraints. A sound mapping between the LTL_f formula and a corresponding BT is then presented, where the BT is structured to use Postcondition-Precondition-Action (PPA) structures. The PPA-style structure allows action nodes to be implemented with off-the-shelf planners. The paper demonstrates the usefulness of the resulting BTs using two case studies. First, the compatibility of a planner's objective with an LTL_f goal is explored (a) by designing plans using Markov Decision Process (MDP)-based planners, (b) by constructing a simple sampling algorithm, and (c) then comparing how well the outcome of the planners match the LTL_f formula. Second, the LTL_f-to-BT algorithm is used to construct *Fetch* robot behaviors that successfully perform *key-door* tasks when perturbations occur.

2   Related Work

LTL [23] is expressive enough to describe complex requirements (safety, liveness, goal-satisfaction) for discrete-time state transition systems. Bacchus et al. [3] were the first to show that linear temporal logic not only can be used to specify system properties but also can be used to specify goals for formal logic systems.

Verifying and synthesizing plans to satisfy complex goal specifications can be computationally prohibitive [17,24], but plans for restricted goal specifications can be found in polynomial time [22]. Thus, there appears to be a trade-off between the expressivity of the specification and the efficiency of the planners. One way to address this expressivity-computability trade-off is by decomposing expressive specifications into small modular pieces. Decomposing complex plans into smaller pieces is not new. Most earlier methods applied decomposition not to the actual specifications but to supporting algorithms ranging from parsing to planning, including reinforcement learning-based approaches [4,19,21,25,31]. Colledanchise et al. [9] were the first to demonstrate direct decomposition of LTL specifications. Vazquez et al. [36] went in a reverse direction by demonstrating a method to construct temporal LTL specifications from a grammar containing atomic specifications.

Frequently, decomposing a LTL specification and creating planners are done concurrently. Generally, the planning process involves: a) converting LTL goal specification to an automaton, b) creating an automaton modeling the environment, c) constructing a product automaton, d) playing the Rabin game using game theory concepts, and e) discretizing the plans [2,5,13,14,16,28,32,33]. Probabilistic Computation Tree Logic is an alternative sometimes used when the planning process is computationally expensive and when uncertainties are present [12,18]. When the guarantees provided by automaton-based planning are not required, sampling-based motion planning algorithms can be used [1,35].

Table 1. BT node type notation.

Node	Figure	Symbol	Success	Failure
Sequence	$\rightarrow$	σ	all children succeed	one child fails
Parallel	$\Rightarrow$	π	all children succeed	one child fails
Selector	?	λ	one child succeeds	all children fail
Decorator	$\Diamond$	δ	User defined	
Action	$\square$	α	task complete	task failed
Condition	$\bigcirc$	κ	true	false

Interestingly, the acceptance of the trace by the automaton can be used as a reward function in some MDP problems [27].

BT representations are (i) equivalent to Control Hybrid Dynamical Systems and Hierarchical Finite State Machines [20] and (ii) generalizations to the Teleo-Reactive paradigm and And-Or-Trees [10]. BT modularity can be combined with the verification properties of LTL. Biggar et al. [6,7] developed a mathematical framework in LTL is used to verify BT correctness without compromising valuable BT traits: modularity, flexibility, and reusability.

3 Behavior Tree and LTL_f Semantics

Assume the world is represented as a finite set of atomic propositions denoted by AP. Assume further that the set of atomic propositions is partitioned into those internal to the behavior tree, AP_b, and those external to the behavior tree, AP_w where the subscripts b and w represents "behavior tree" and "world", respectively. Thus, $AP = AP_b \cup AP_w$.

Let $\{null, a_1, a_2, \ldots, a_m\} \in A$ denote the set of valid actions, where $null$ indicates inaction by the agent. The world is "open", which means that atomic propositions can change even when the agent does not act. Inaction is represented by $null$. Assume a discrete time world where changes occur at fixed intervals. Further, assume uncertainty in the world's starting state, and assume that the world is deterministic given its starting state. Differences in a state trajectory induced by a fixed action sequence are modeled as random selection of starting state (Table 1).

3.1 Behavior Tree Semantics

A *Behavior Tree* (BT) is a directed rooted tree where the internal nodes are called *control nodes*, and the leaf nodes are called *execution nodes*. A node can be in only one of three states: *running* (processing is ongoing), *success* (the node has achieved its objective), or *failure* (anything else). This subsection presents a formal syntax and semantics for behavior tree behavior.

We define the standard two types of execution nodes, (*Action* and *Condition*), but we only define three of the four standard types of control nodes: *Sequence*,

Selector, and *Decorator* [11]. The *Parallel* node type is not implemented, which has some consequences for how quickly missions can execute (discussed in the next section). The syntax of the nodes define the structure of the BT. We use N to denote a node of arbitrary type and let the syntax make clear the actual type in the semantics. Node syntax uses prefix notation, (N argument_list), where the node operation type is followed by the arguments on which the node operates. For example, (σ N^L N^R) represents a sequence node (σ) that operates on left and right child nodes (N^L and N^R, respectively).

An action node, (α id) where id is a unique identifier, executes the action or plan that is referred to by id in the world. A condition node, (κ ϕ) where ϕ is a quantifier free first order logic formula, evaluates some property of the world. A sequence node, (σ N^L N^R), is a logical conjunction with *short-circuit* semantics. Short-circuit semantics evaluate operands in order stopping when the expression outcome can be logically concluded. For simplicity, we restrict ourselves to at most two operand nodes, a left node and right node, in the sequence and other similar node types. This restriction does not affect expressiveness. A selector node (λ N_0 N_1) acts as a disjunctive proposition with short-circuit semantics. A decorator node, (δ id N), follows a user defined rule identified by "id" that decides its status from its child.

BT semantics are defined recursively over the syntax of the tree with a *tick* function. A *tick* is recursively sent from a controller, which is external to the behavior tree, to the root tree node, which passes the *tick* to its children using a depth-first-search (DFS) tree traversal from left to right.

In the BT semantics, system state and dynamics are abstracted to be a function s : $AP \mapsto \{success, failure, running\}$ mapping atomic propositions to analogues for true, false, and unknown respectively. We omit the details of the ternary propositional logic to not distract from the presentation. Only decorator nodes keep an internal state, so the set AP_b contains the internal states of all decorator nodes in the behavior tree.

For many roboticists and agent designers, defining state as a function differs from the conventional approach as defining it as a vector of state variables. We use the function-based representation of state because it makes expressing behavior tree semantics considerably easier than using a state vector-based representation. Note that the vector-based representation can be obtained by assigning an order to elements of $AP = \{p_1, p_2, \ldots\}$ representing state $s = [s(p_1), s(p_2), \ldots]$ as vector of atomic proposition values returned by the state function s.

Each BT node returns two things: its status (b) and the updated state function (s), yielding the tuple (b, s). (Returning the updated state function is equivalent to returning a "next state vector".) Only decorator nodes update internal behavior tree state and only action nodes update external world state. Behavior tree semantics are defined recursively, with two base cases: action nodes, Eq. (1), and condition nodes, Eq. (2). For condition nodes, we indicate with $[\![\phi]\!]_s$ the evaluation of the propositional formula ϕ in the logic given the state s. The notation ($[\![\phi]\!]_s$ s) can be read as "the operator $[\![\phi]\!]_s$ gives meaning to state s." The L and R superscripts indicate the left and right children, respectively, of the sequence and selector nodes.

$$tick \ s \ (\alpha \ id) = run \ s \ id \tag{1}$$

$$tick \ s \ (\kappa \ \phi) = ([\![\phi]\!]_s \ s) \tag{2}$$

$$tick \ s \ (\sigma \ N^L \ N^R) = \begin{cases} (b^L, s^L) & \textbf{if } b^L \in \{fail, running\} \\ & \text{where } (b^L, s^L) = tick \ s \ N^L \\ tick \ s^L \ N^R \ \textbf{otherwise} \end{cases} \tag{3}$$

$$tick \ s \ (\lambda \ N^L \ N^R) = \begin{cases} (b^L, s^L) & \textbf{if } b^L \in \{success, running\} \\ & \text{where } (b^L, s^L) = tick \ s \ N^L \\ tick \ s^L \ N^R \ \textbf{otherwise} \end{cases} \tag{4}$$

$$tick \ s \ (\delta \ id \ N) = compute \ s \ id \ N \tag{5}$$

Actions, (1), rely on an external *run* function to execute the indicated plan from the given state. That function results in a new updated state that is propagated downstream (e.g., from leaf nodes to parent nodes) in the semantics. Sequences, (3), and selectors, (4), *tick* the second child depending on the return status of the *tick* on the first child according to their respective semantics. Decorators, (5), rely on an external *compute* to execute the indicated rule that decides whether or not the child sees the *tick* and what the return status should be. Like the *run* function, *compute* may result in a new updated state that is propagated downstream in the semantics.

Let $\mathbb{T}$ denote a specific behavior tree. The language of $\mathbb{T}$ is all possible sequences resulting from some set of valid initial states. To define this language, we assume a controller function that calls *tick* at a fixed frequency, passes *tick* to the root note, and repeats until the tree resolves to a final end state. We identify the tree as a controller function, $(b, [s_0, s_1, \ldots]) = \mathbb{T}(s_0)$, where the subscripts associated with the states represent changes in state. The first tick always uses the starting state s_0 with subsequent ticks using the resulting state from the previous tick. The tree returns the final state of the tree ($b \in \{success, fail\}$) and the finite sequence of states observed after each call to *run* and *compute* while ticking the tree ($[s_0, s_1, \ldots]$).

We make three assumptions about the relationship between the behavior tree and the external world:

- the *run* and *compute* functions eventually succeed or fail,
- the tick frequency is such that the previous tick on the tree always returns before the next tick, and
- the tick frequency is fast enough that external changes in the world (e.g., changes not caused by *run*) are observed in the tree.

We now define the *traces* and the *language* of a BT.

Definition 1 (Traces). *The set of traces generated by $\mathbb{T}$ given a set of initial states S_0 is*

$$\Gamma(S_0, \mathbb{T}) = \big\{ [s_0, s_1, \ldots] \mid s_0 \in S_0 \wedge (b, [s_0, s_1, \ldots]) = \mathbb{T}(s_0) \wedge$$
$$(b == success \ \vee \ b == failure) \big\}$$

where the trace only includes external states from AP_w *and omits internal states from* AP_b.

Observe that the notation $[s_0, s_1, \ldots]$ represents a finite trace with one or more states. Traces are finite because the tree terminates with success or failure. Also observe that the tree is deterministic and that traces are defined with respect to a set of starting states S_0. The set of starting states models nondeterminism as uncertainty over starting states, meaning that a starting state is chosen nondeterministically and the trace that results is deterministic.

Stated simply, the traces make up the set of all external state sequences that can be produced by a BT when the BT resolves to success or fail. We define the *language* of a BT as the subset of trajectories that resolve to success.

Definition 2 (Language of a BT). *The language of a BT* $\mathbb{T}$ *given a set of initial states* S_0 *is*

$$L(S_0, \mathbb{T}) = \big\{ [s_0, s_1, \ldots] \mid s_0 \in S_0 \wedge (b, [s_0, s_1, \ldots]) = \mathbb{T}(s_0) \ \wedge$$
$$b == \text{success} \big\} \tag{6}$$

The remainder of the presentation omits S_0 from the notation preferring $L(\mathbb{T})$ with the implicit assumption that it is defined over some set of initial world and decorator node states.

3.2 *LTL$_f$* Semantics

LTL$_f$ formulas are constructed from a finite set of propositional variables drawn from the set of atomic propositions, AP combined using *logical operators* and *temporal modal operators*. Conventional logical operators are used: $\neg$, $\vee$, and $\wedge$. The temporal modal operators are **X** ("neXt"), **U** ("Until"), **F** ("Finally", meaning at some time in the future), and **G** ("Globally"). Unary operators take precedence over binary, the **U**ntil operator takes precedence over the $\wedge$ and $\vee$ operators, and all operators are left associative.

A state is defined as a subset of atomic propositions that are true at time i, $s_i \subseteq AP$. *LTL$_f$* formulas operate on a trace, denoted by $\tau = [s_0, s_1, \ldots]$ that consists of a sequence of states. Trace segments are indexed using the following notation: $\tau[i] = \mathbf{s}_i$ and $\tau[i : j] = [\mathbf{s}_i, \ldots, \mathbf{s}_j]$. Let m denote the maximum length of a finite *LTL$_f$* formula. Thus, a full trace is $\tau[0 : m]$ and the "suffix" of a trace beginning at time i in $0 < i \leq m$ is $\tau[i : m]$.

The semantic interpretation of an *LTL$_f$* formula, ψ, is given using the *satisfaction relation*, $\models$, which defines when a trace satisfies the formula. A trace suffix $\tau[i : m]$ satisfying an *LTL$_f$* formula ψ is denoted using prefix notation by $\models \tau, i \ \psi$ and is inductively defined (using the $\equiv$ to indicate "defined") as follows:

- $\models \tau, i \; A \equiv A \in \tau[i]$ and $A \equiv$ **true**
- $\models \tau, i \; (\neg \psi) \equiv (\not\models \tau, i \; \psi)$
- $\models \tau, i \; (\wedge \; \psi_1 \; \psi_2) \equiv \models \tau, i \; \psi_1$ and $\models \tau, i \; \psi_2$.
- $\models \tau, i \; (\vee \; \psi_1 \; \psi_2) \equiv \models \tau, i \; \psi_1$ or $\models \tau, i \; \psi_2$.
- $\models \tau, i \; (\mathbf{X} \; \psi) \equiv \models \tau, (i + 1) \; \psi$.
- $\models \tau, i \; (\psi_1 \; \mathbf{U} \; \psi_2) \equiv \models \tau, k \; \psi_1$ and $\models \tau, j \; \psi_2$ where $\exists j \in \{i, i + 1, \ldots, m\}$ such that $\forall k \in \{i, i + 1, \ldots, j - 1\}$.
- $\models \tau, i \; (\mathbf{G} \; \psi) \equiv \models \tau, k \; \psi$ where $\forall k \in [i, m]$.
- $\models \tau, i \; (\mathbf{F} \; \psi) \equiv \models \tau, k \; \psi$ where $\exists k \in [i, m]$.

Observe that traces in LTL_f are made up of boolean elements not the ternary elements used in behavior trees. The following section constructs a specific LTL_f formula that is satisfied by traces generated by a formula-specific BT whenever the behavior tree returns success. A key step is mapping between traces generated by a successful BT and a trace evaluated by an LTL_f formula.

4   Constructing a Behavior Tree from an LTL_f Task Formula

This paper restricts attention to goal specifications to *task* structures that follow a postcondition, precondition, and action (PPA) structure. PPA structures check the postcondition before trying an action, and have proven useful for agent-based and robotics applications [11]. The resulting subset of LTL_f that uses the PPA structure is called *PPA-LTLf*. This section shows that a BT can be constructed such that the traces generated by the BT satisfy a formula expressed using the *PPA-LTLf* structure.

4.1   Task Formula and Task Behavior Tree

PPA-Style Task Formula. The basic structure of a PPA-structured achievement goal can be represented using the propositional logic formula $PPATask = \vee\{PoC\{\wedge[PrC][Action]\}\}$ where *PoC*, *PrC*, and *Action* are postcondition, precondition, and action propositions, respectively. The *Action* proposition is true iff and only if the action satisfies the post-condition, so the formula could have been rewritten as $PPATask = \vee\{PoC\{\wedge[PrC][PoC]\}\}$. If execution of the expression is performed from left to right the $\vee$ operator means that the action will not be executed if the postcondition is already satisfied. If the postcondition is not satisfied, then the $\wedge$ operator ensures that action is only executed when the precondition is satisfied. To help extend the formula to more general conditions, the two operands of the $\vee$ operator are delineated with curly braces and the two operands of the $\wedge$ operator with square braces.

A more general goal includes global and task constraints,

$$\psi = \wedge\Big\{\mathbf{G} \; GC\Big\}\Big\{ \vee \; \big[PoC\big] \; \Big[\wedge \; \big(PrC\big) \; \big(\mathbf{U} \; (TC) \; (Action)\big)\Big]\Big\} \qquad (7)$$

$$= \wedge\Big\{\mathbf{G} \; GC\Big\}\Big\{ \vee \; \big[PoC\big] \; \Big[\wedge \; \big(PrC\big) \; \big(\mathbf{U} \; (TC) \; (\wedge \; PoC \; GC)\big)\Big]\Big\}$$

where PoC, PrC, GC, and TC are boolean formulas that encode task postconditions, task preconditions, global constraints, and task constraints propositions, respectively. The two forms of the equation emphasize that an *Action* is satisfied if and only if both the post condition and the global constraint are satisfied. Large curly braces delineate the two operands of the leftmost ∧ operator, which requires that the global constraint is always satisfied *and* the postcondition is eventually satisfied. Large square braces delineate the two operators of the ∨ operator where the first operand represents the postcondition and the second operand represents that precondition/action pair. The simple postcondition has been replaced with a check that the global constraint and postcondition are simultaneously satisfied. The precondition is delineated by the large parentheses. The action has been replaced by an until operator, which says that the task constraint must hold until the action has satisfied the postcondition and global constraint. This formula can be simplified since there are redundant conditions (e.g., the requirement that the global constraint be satisfied globally and the requirement that the action satisfy the global constraint), but this form allows a direct mapping onto a behavior tree.

BT for a PPA-Style Task. Figure 1 shows the *task behavior tree*, denoted by $\mathbb{T}_\psi$, constructed from the task formula ψ in Eq. (7). Let $L(\psi)$ denote the set of all traces that satisfy ψ. We will show that the behavior tree is a sound implementation of the task formula, which means that $L(\mathbb{T}_\psi) \subseteq L(\psi)$.

We now describe how the components of the task formula are implemented in the BT. We do this by describing each node as we encounter it while performing a depth first search tree traversal of the tree. We provide a description of how signals flow between components of the tree, relating the signals to the BT semantics in Eqs. (1)–(5). The edges in the tree are labeled with the signals that correspond to a successful execution of the BT. The downward curved arrows represent states passed to children and the upward curved arrows represent the return statuses and states passed to parents. The subscripts indicate changes in state. The return status of *success* and *failure* are denoted in shorthand form by $b = c$ or $b = f$, respectively.

Controller. The downward arrow at the top of the tree represents the controller that generates ticks at a fixed frequency. The s_0 label next to the arrow is the initial state.

Sequence Node σ_{PPATask}. This sequence node implements the leftmost *and* operator $\wedge\{\mathbf{G}\ GC\}\{\cdot\}$ of Eq. (7). Its left child is the condition node κ_{GC} that evaluates the global condition ($[\![GC]\!]_s\ s$) from Eq. (2). This condition node receives state s_0 from its parent and returns success c, indicating that the global constraint was satisfied. The condition node does not modify state, so it returns s_0.

Selector Node λ_{PocBlk}. This selector node implements the *or* operator $\vee\ [PoC]\ [\cdot]$ of Eq. (7). This selector node receives the state s_0 from its parent and returns a success status (c) to its parent along with the state modified

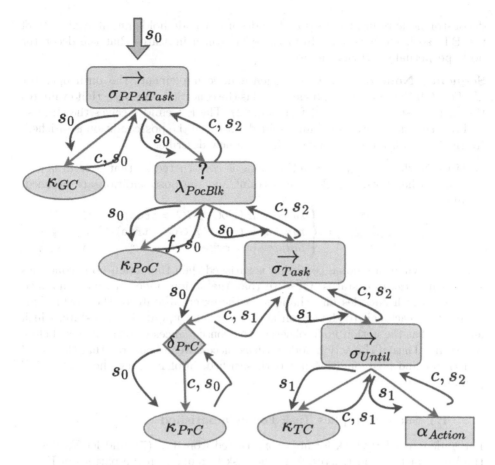

Fig. 1. Task BT $\mathbb{T}_\psi$ constructed for PPATask formula ψ.

by its descendants. Its left child is the condition node κ_{PoC} that evaluates the postcondition ($[\![PoC]\!]_s$ s) from Eq. (2). The signal labels indicate that the child returns an unmodified state s_0 and a failure status f. The failure status was chosen to illustrate a task that is successful because action is taken in the world.

Sequence Node σ_{Task}. This sequence node implements the *and* portion $\wedge$ (PrC) ($\cdot$) of Eq. (7). It's left child checks the precondition and its right child implements the until operator. We describe each child separately.

Decorator Node δ_{PrC}. The decorator node and its child check the precondition. The precondition needs to be satisfied only during the tick in which the action begins to execute. Thus, the decorator node always returns *success* if the precondition is satisfied, blocking the need to recheck the precondition. Its child is the condition node κ_{PrC} that evaluates the precondition ($[\![PrC]\!]_s$ s) from Eq. (2). The precondition node receives the state s_0, and it returns success c (indicating that the precondition was satisfied) and an unmodified state s_0. The decorator node returns success c. The decorator node holds an internal state of the BT, so it changes state from s_0 to s_1, which indicates that the decorator node perpetually returns success.

Sequence Node σ_{Until}. This sequence node implements the until operator $\mathbf{U}\,(TC)\,(Action)$ in Eq. (7). Its left child is the condition node κ_{TC} that evaluates the task constraint ($[\![TC]\!]_s\,s$) from Eq. (2). The left child receives the state s_1 and returns success c and the unmodified state s_1 if the postcondition is satisfied. Its right child is an action node, which we now describe.

Action Node α_{Action}. The action node is constructed so that it returns success only when both the global constraint and the postcondition are satisfied. Formally,

$$\alpha_{Action}[t] = \begin{cases} success & \text{if } (PoC \wedge GC = \mathbf{true}) \\ running & \text{if } (\neg PoC \wedge GC = \mathbf{true}) \\ failure & \text{otherwise} \end{cases}$$

Note that while the global condition is satisfied then the action node continues to return a *running* status. Observe that the action node returns a modified state s_2, which encodes any changes to the external state of the world. This means that each call to the action node returns an updated world state, which assumes that the action node observes the consequences of what it does before returning. Thus, if the action node returns success we are ensured that the global constraint and postcondition are both satisfied, implementing the $\wedge\ PoC\ GC$ portion of Eq. (7).

4.2 BT Success Implies Task Formula Satisfied

Let ψ denote a PPATask formula constructed from Eq. (7), and let $\mathbb{T}_\psi$ denote the behavior tree constructed from the task formula ψ in the pattern of Fig. 1. We can define the language of a task formula analogously to the language of a BT.

Definition 3 (Language of a Task Formula). *The language of an LTL_f task formula constructed from Eq. (7) is defined as*

$$L(\psi) = \{\tau : \tau \text{ is a trace that satisfies } \psi\}$$

The following theorem states that the task behavior tree is a *safe substitution* of the LTL_f formula. This means that the formula is an over approximation of the BT so any property we conclude in the formula we can conclude about the BT. Before stating and proving the theorem, recall from Definition 1 that the internal states of the BT are excluded from the definition of BT traces.

Theorem 1. *Given a task formula, ψ, and the behavior tree constructed from the task formula, $\mathbb{T}_\psi$, $L(\mathbb{T}_\psi) \subseteq L(\psi)$*

Proof. The proof has two cases.

Case 1: There is only one element of the trace. Each call to the action node α_{Action} appends an element to the trace, so a successful trace with a single element produced by $\mathbb{T}(\psi)$ never reaches α_{Action}. Since Eq. (1) says that a successful trace requires one of the children of λ_{PocBlk} to return success, this case requires the left child of λ_{PocBlk} to return success, which can only occur when the postcondition is satisfied. And since Eq. (3) says that $\sigma_{PPATask}$ requires both of its children to return success, a successful trace with a single element must satisfy both PoC and GC. Applying to the task formula in Eq. (7), ψ is satisfied because both GC and PoC are satisfied.

Case 2: There is more than one element of the trace. Let the elements of the trace be denoted by $\tau = [\tau_0, \tau_1, \ldots, \tau_k]$, where the trace succeeds on the kth state change. The first trace element τ_0 cannot satisfy both GC and PoC because otherwise Case 1 would have applied. However, the precondition must have been satisfied by the initial element of the trace because otherwise the decorator node would have returned failure and the action node would never have been called. Thus the PrC component of Eq. (7) is satisfied. The task constraint must have been satisfied for all ticks $0, 1, \ldots, k - 1$ since otherwise the left child of σ_{Until} would have returned failure. Similarly, the global constraint must have been satisfied for all ticks $0, 1, \ldots, k - 1$ since otherwise Eq. (2) says that κ_{GC} would have returned failure, causing $\sigma_{PPATask}$ to return failure. Thus, both TC and GC were satisfied until tick $k - 1$. The action node appends τ_k to the trace when it returns success, and this can only occur when the action node observes that both the postcondition and global constraint were satisfied. This means that τ_k satisfies PoC and GC. Consequently, the $\mathbf{U}(TC) (\wedge PoC\ GC)$ portion of Eq. (7) is satisfied, which means that ψ is satisfied.

In both cases, $\tau \in L(\mathbb{T}_\psi)$ implies $\tau \in L(\psi)$.   □

4.3 Empirical Verification

It builds confidence to empirically verify that any successful BT execution produces a trace that satisfies the task formula. We simplify the problem by assuming that the constraints and conditions, $\{PoC, PrC, GC, TC\}$ are single propositions, yielding $2^4 = 16$ possible relevant possibilities. A simulation environment was created that caused random transitions between the proposition variables. Each simulation randomly assigned a starting state. Traces of length five were evaluated, yielding a set 16^5 possible unique traces. A BT was created using the PyTrees [30]. The environment has a *step* method that implements the *tick*.

The task formula was implemented using the python *flloat* library [15]. We ran 2^{20} independent trials with random traces, and measured both when the BT returned success or failure and when the trace was satisfied or not.

The confusion matrix from the experiments is shown below. Note that for all 26.81% successful trace generate by the BT, all of them were valid traces. The status of BT and LTLf parse are in complete agreement for 2^{20} random traces (Table 2).

5 BT for LTL_f Task-Based Missions

This section defines a mission as a temporal sequence of PPA-style tasks constructed from a subset of LTL_f. The set of possible missions is defined using a

Table 2. Confusion matrix of PPATask BT experiments.

		BT Status	
		Success	Failure
LTL_f Satisfied	True	26.8%	0
	False	0	73.2%

context free grammar. A behavior tree is then constructed from the grammar. Since the PPATask BT is a safe substitution for the LTL_f formula, a mission can be written in LTL_f and then implemented directly with the PPATasks tree without risking violating the LTL_f formula. We use this safe substitution property to prove by induction that every successful mission generated by the behavior tree satisfies the LTL_f formula describing the mission. The result of this process is the construction of a behavior tree that is a *safe substitution* for the corresponding mission LTL_f formula.

5.1 Constructing a Mission Behavior

The temporal sequence is specified using LTL_f operators, but not all temporal sequences are permitted. Combining PPA-style tasks using the LTL operators is constrained by both (a) the nature of tasks and (b) the task BT implementation of a PPA-style task. This subsection presents the mission grammar, describes what can and cannot be done using the mission grammar, and constructs a behavior tree for a mission LTL_f formula.

Mission Grammar. The mission grammar is expressed in Productions (M1)–(M6)) using prefix operator notation. Productions (M1) and (M3) enforce precedence, which from highest to lowest is **Finally**, **Until**, and $\lor$. Productions (M2) and (M4) enforce the left associativity required of the binary LTL operators. Production (M6) means that a Mission is composed of one or more PPA-style tasks. The parentheses in Productions (M5)–(M7) ensure that tasks in a multi-task mission are distinct.

$$\langle \text{Mission} \rangle :: = \langle \text{L1} \rangle \tag{M1}$$

$$:: = \lor \langle \text{Mission} \rangle \langle \text{L1} \rangle \tag{M2}$$

$$\langle \text{L1} \rangle :: = \langle \text{L2} \rangle \tag{M3}$$

$$:: = \mathbf{U} \langle \text{L1} \rangle \langle \text{L2} \rangle \tag{M4}$$

$$\langle \text{L2} \rangle :: = \mathbf{F} \left(\langle \text{Mission} \rangle \right) \tag{M5}$$

$$:: = \left(\langle \text{PPATask} \rangle \right) \tag{M6}$$

Mission Assumptions and Restrictions. First, the mission grammar does not allow precise task timing. This can be seen by the absence of both the ∧ and NeXt operators from LTL_f. Semantically, ∧ means simultaneous completion time and *NeXt* means completion on the next state change. If precise timing is required between two tasks, these tasks must be combined in a single task action node. Instead of precise timing, a mission implements task sequencing using **Finally** and **Until**.

Second, missions are not optimized for speed. The short circuit semantics for a ∨ operator mean that the first operand is evaluated before the second. Parallel computations could potentially be much faster when this is implemented on a behavior tree, but this would require that the two tasks avoid side affects like competing for the same resource, occupying the same physical location, or inducing a race condition. Future work should specify precise conditions for what it means for two tasks to be independent.

Constructing the Mission BT. Let Ψ (capital ψ) denote a mission formula derived from the grammar above and expressed in LTL_f form. The mission BT $\mathbb{T}_\Psi$ is constructed from the expression tree in the following manner.

Leaf Nodes of $\mathbb{T}_\Psi$. Each leaf node is represented by a task tree $\mathbb{T}_\psi$, effectually enforcing the parentheses in Production (M6).

∨ **Operator.** Production (M2) says that some missions can be accomplished in multiple ways, *either* by performing the combination of tasks in the tree descending from the left branch *or* the right branch. The right-hand side of Production (M2) is ∨⟨Mission⟩⟨L1⟩, which is implemented as a subtree rooted at a selector node, $\lambda_{\langle Mission\rangle}$ as shown in Fig. 2.

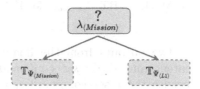

Fig. 2. Mission GBT $\mathbb{T}_{\langle Mission\rangle}$ where ⟨Mission⟩ :: = ∨⟨Mission⟩⟨L1⟩.

Until Operator. Production (M4) says that some missions require (a) the task determined by applying productions rooted at ⟨L2⟩ to succeed (b) *until* the task determined by applying productions rooted at ⟨L3⟩ succeeds. Figure 3 shows how the until operator in this production, U⟨L2⟩⟨L3⟩, is implemented as a subtree rooted at a sequence node type, σ_U. The sequence node ensures that the left most subtree succeeds before the right subtree is executed.

Finally Operator. Production (M6) uses the finally operator. The semantics of the finally operator say that if the child subtree, $T_{\langle Mission\rangle}$, fails then it should be given another chance to succeed. The finally operator is implemented using

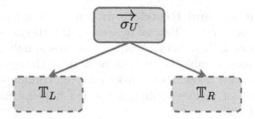

Fig. 3. Mission BT $\mathbb{T}_{\langle \text{Mission} \rangle}$ for **U** $\Psi_{\langle \text{L1} \rangle}$ $\Psi_{\langle \text{L2} \rangle}$.

a decorator node, δ_F, that returns success as soon as its child subtree $\mathbb{T}_{\langle \text{Mission} \rangle}$ returns success. If the child subtree $\mathbb{T}_{\langle \text{Mission} \rangle}$ returns failure, then the decorator node returns running and resets the internal states (i.e., the decorator nodes) of all descendant task trees.

Note that the δ_F decorator node separates out the subtree descending from the $\langle \text{Mission} \rangle$ nonterminal from the rest of the BT, effactually enforcing the parentheses in Production (M5) (Fig. 4).

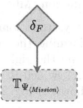

Fig. 4. Mission GBT $\mathbb{T}_{\langle \text{Mission} \rangle}$ for **F** Ψ.

In summary, there are two outputs from the finally decorator node: the status returned to its parent and the *reset* command to the task tree if $\mathbb{T}_{\langle \text{Mission} \rangle}[t] = \textit{failure}$. Because the reset command can cause infinite retries, we include a timeout condition if execution time meets a threshold T_θ. The decorator behavior is defined as

$$\delta_F[t] = \begin{cases} \textit{success} & \text{if } \mathbb{T}_{\langle \text{Mission} \rangle}[t] = \textit{success} \\ \textit{failure} & \text{if } (t \geq T_\theta) \\ \textit{running} & \text{otherwise} \end{cases}$$

Mission BT Semantics. Mission semantics are expressed in terms of PPATask success or failure. We omit a precise discussion of mission semantics in the interest of space, but we briefly describe the semantics of a mission tree here. Each mission BT has a controller that issues *tick* to the root node of the tree. This *tick* is passed down the tree in a left-to-right depth first search traversal. The leaf nodes of a mission tree are composed of PPA Tasks, which are themselves

BTs. Like with the task trees, the BT implementations of the binary operators used in a mission behavior tree, $\vee$ from M2 and the $\mathbf{U}$ from M4, do not evaluate their right child until their left child resolves to success or failure. Consequently, leaf nodes are executed from left to right in the tree and, except for the leftmost leaf node, no leaf node is executed until all leaf nodes to its left resolve to success or failure.

5.2   BT Success Implies Formula Satisfied

The *mission trace*, denoted τ, is initialized as the trace produced by the leftmost PPATask leaf node. When the next leaf node resolves to success or failure, the trace from that PPATask BT is appended to the mission trace, and so on.

Define the languages for a mission LTL_f formula Ψ and a mission BT $\mathbb{T}_\Psi$, respectively as follows:

- $L(\Psi) = \{\tau : \tau \text{ is a mission trace that satisfies } \Psi\}$
- $L(\mathbb{T}_\Psi) = \{\tau \colon \tau \text{ is a mission trace generated when } \mathbb{T}_\Psi = success\}$

Theorem 2. $L(\mathbb{T}_\Psi) \subseteq L(\Psi)$

Proof. The proof is by induction, with multiple induction steps, one for each LTL_f operator appearing in the mission grammar.

Leaf Nodes. The base case is for $\Psi = \psi$, a mission consisting of a single PPA-style task. The base case is proven in Theorem 1.

Subtrees Rooted in the Finally Operator. Consider a subtree rooted at $\mathbf{F}\left(\langle\text{Mission}\rangle\right)$, and let $\mathbb{T}_{\langle\text{Mission}\rangle}$ denote the subtree that descends from the δ_F decorator node. The induction hypothesis is that $\mathbb{T}_{\langle\text{Mission}\rangle}$ returns success implies the mission formula $\Psi_{\langle\text{Mission}\rangle}$ is satisfied. The finally operator can only return success if $\mathbb{T}_{\langle\text{Mission}\rangle}$ returns success, even if prior runs of $\mathbb{T}_{\langle\text{Mission}\rangle}$ fail. This means that the trace produced by $\mathbb{T}_{\langle\text{Mission}\rangle}$ eventually satisfies $\Psi_{\langle\text{Mission}\rangle}$, which means that $\mathbf{F}\Psi_{\langle\text{Mission}\rangle}$ is satisfied.

Induction Step for the Until Operator. The second induction step is for production (M4), which corresponds to the formula $\mathbf{U}\ \Psi_L\ \Psi_R$, where L and R indicate the formulas encoded in the left and right children of the Until sequence node. The induction hypothesis has two parts: $\mathbb{T}_{\Psi_L} = $ success implies that Ψ_L is satisfied, and $\mathbb{T}_{\Psi_R} = $ success implies that Ψ_R is satisfied. The subtree implementing the until operator, $\mathbb{T}_{\mathbf{U}\ \Psi_L\ \Psi_R}$ is rooted at a sequence node with left and right children, $\mathbb{T}_{\Psi_L}$ and $\mathbb{R}_{\Psi_R}$, respectively. The sequence node returns success only when $\mathbb{T}_{\Psi_L} = $ success up to the time when $\mathbb{T}_{\Psi_R}$ returns success. The fact that $\mathbb{T}_{\Psi_L} = $ success continues to return success while $\mathbb{T}_{\Psi_R}$ executes means that the LTL_f formula corresponding to the $\langle\text{L1}\rangle$ portion of the production $\mathbf{U}\langle\text{L1}\rangle\langle\text{L2}\rangle$ is satisfied for all atomic states up to the time that the $\mathbb{T}_{\Psi_R}$ returns success. When $\mathbb{T}_{\Psi_R}$ returns success, the LTL_f formula corresponding to the $\langle\text{L2}\rangle$ portion of the production is satisfied. Thus, the until operator is satisfied.

Induction Step for the Or Operator. The proof for the $\vee$ operator is omitted since it follows a similar pattern as the proof for the $\mathbf{U}$ operator, evaluating the trace generated by the left and right children sequentially. $\square$

6   Planner-Goal Alignment Example

Given the BT produced from a given mission formula, it is necessary to choose a plan or policy for each action node. Traditional LTL plan synthesis using advanced automata (Rabin and Buchi) requires a suitable environment model and is of high computation complexity. When the goal specifications are valid but sufficient information about the environment is unknown, traditional automata-based solutions are infeasible. However, the BT does not specify what type of planner is required. Instead, off-the-shelf planners can be used to design the plans or policies used in the BT's action nodes. This is demonstrated using two planners: policy iteration, which uses reward functions to represent goals that may or may not align with the LTL_f formula, and a state-action table that uses the BT return status to update the probability of actions in successful or unsuccessful action sequences.

6.1   Problem Formulation

The *Mouse and Cheese* problem is a classic grid world problem [26]. This paper uses a 4x4 grid giving sixteen world locations indexed by $s_{j,k}$. There is one atomic proposition $A_{j,k}$ for each state, where $A_{j,k} = true$ indicates that the mouse occupies location $s_{j,k}$. The mouse may not occupy two locations simultaneously, $A_{i,j} == true \rightarrow A_{k,\ell} == false$. Location $s_{4,4}$ contains the cheese, and the mouse automatically picks up the cheese when it occupies that cell. An atomic proposition $Cheese = true$ indicates that the mouse has the cheese. Location $s_{4,2}$ is dangerous, denoted by atomic proposition $Fire$, and the mouse should avoid this state. Atomic proposition $Home$ indicates whether the agent is at home location $s_{3,1}$. The state vector at time t is the vector of truth values for each atomic proposition $\mathbf{s}_t = [s_{1,1}, s_{1,2}, \ldots, s_{4,4}, Cheese, Fire, Home]$.

Paraphrasing from [26], the agent has four actions: move up, left, right, or down. if the agent bumps into a wall, it stays in the same square. The agent's actions are unreliable, i.e., the "intended" action occurs with some probability p_{in} but with some lower probability, agents move at the right angles to the intended direction, $1 - p_{in}$. The problem gets harder for the agent as p_{in} decreases.

The mouse and cheese problem is a sequential planning problem, requiring the agent to find the cheese and then return it to the home location. This is an achievement goal that requires two separate plans (or policies): (a) find the cheese avoiding fire and (b) return home avoiding fire after finding the cheese. There are many planners that can solve this problem (e.g., MAX-Q and other hierarchical learners), and the point of this section is not to argue for the best way to solve the problem. Rather, the point is to explore how well different types of planners, one for each task, align with the overall mission goal.

The goal of the mouse is to retrieve the cheese while avoiding the fire location and getting back at the home location. An LTL_f formula from the mission grammar for the sequential find-the-cheese-and-return home (C2H) and the corresponding PPA task formulas from Eq. (7) are

$$\Psi^{C2H} = \mathbf{U} \; \mathbf{F}\psi^{\text{Cheese}} \; \mathbf{F}\psi^{\text{Home}}$$
$$\psi^{\text{Cheese}} = \vee(\wedge\neg Fire \; Cheese)(\wedge(\wedge\neg Fire \; True)$$
$$(\mathbf{U} \; True \; \wedge \; Action_{\text{Cheese}}))$$
$$\psi^{\text{Home}} = \vee(\wedge\neg Fire \; Home)(\wedge(\wedge\neg Fire \; Cheese)$$
$$(\mathbf{U} \; True \; \wedge \; Action_{\text{Home}}))$$

The BT action nodes $Action_{\text{Cheese}}$ and $Action_{\text{Home}}$ execute the plans that lead mouse to the cheese and return home, respectively.

6.2   Action-Node Policies via Policy Iteration

This section uses policy iteration to create a policy for $Action_{\text{Cheese}}$ and again to create a policy for $Action_{\text{Home}}$. The reward structure for the cheese task is $\mathbf{r} = (r_{\text{other}}, r_{\text{cheese}}, r_{\text{fire}})$, where the rewards are for occupying any grid cell other than home or fire, having the cheese, and occupying the fire cell, respectively. The reward structure for the home task is $\mathbf{r} = (r_{\text{other}}, r_{\text{home}}, r_{\text{fire}})$, where the rewards are for occupying any grid cell other than home or fire, occupying the home cell, and occupying the fire cell, respectively. Restrict attention to situations where $r_{\text{cheese}} = r_{\text{home}} = r_{\text{good}}$, which allows results to be represented using a triple $\mathbf{r} = (r_{\text{other}}, r_{\text{good}}, r_{\text{fire}})$.

Using policy iteration to create policies from rewards encodes goal in two ways: once in the BTs via the return values of the root node, and once in the reward structures themselves. The reward-goal alignment problem is well-known, and the problem is explicit when the goal is encoded directly in the LTL_f formula but indirectly in the reward structures. By construction of the BT, every successful trace satisfies the goal, but many traces fail or time-out depending on the reward structure.

To illustrate, five hundred twelve independent experiments are conducted for various intended action probabilities p_{in} and rewards values. The dependent variables are *success probability*, which is defined as the number of simulations where Ψ^{C2H} is satisfied divided by the total number of simulations, and *trace length*, which is the length of the trace. Experiments used the following:

Parameter	Values
r_{other}	$\{-1.5, -1.4, -1.3, \ldots, -0.2, -0.1, -0.04\}$
$r_{\text{cheese}} = r_{\text{home}}$	$\{0.1, 0.5, 1.0, 2.0, 5.0, 10.0\}$
r_{fire}	$\{-10, -5, -2, -1, -0.5, -0.1\}$
p_{in}	$\{0.4, 0.45, 0.5, \ldots, 0.9, 0.95\}$

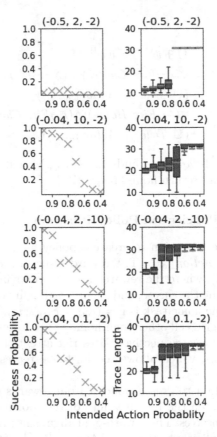

Fig. 5. The left column shows the success probability and the right column shows the trace length with various reward structures and intended action probabilities p_{in}. Higher negative rewards in intermediate states leads to lower mission accomplishment rate.

Figure 5 shows the average success probability from a sample of simulations. Each row represents a distinct reward structure, and the x-axis of each subplot represents intended action probabilities p_{in}. Note that when r_{other} has large negative values, the success probability decreases as the p_{in} decreases, which is consistent with standard results in reinforcement learning. In contrast, the trace length decreases when r_{other} has large negative values because the agent ends up in the fire. The lesson from this experiment is that poor planning in a difficult problem overrides the benefit of the guarantee Ψ^{C2H} is satisfied whenever the BT returns success.

6.3   Action-Node Policies with BT Feedback

The property that every successful trace of the BT satisfies the LTL_f formula from which it was derived can be used to learn policies for the action nodes while the BT is running.

Algorithm. Represent the policy for an action node using a probabilistic state-action mapping, $\pi : S \rightarrow \Delta(A)$, where S is the set of possible states and $\Delta(A)$ is a probability distribution over actions. Thus, the policy is a conditional probability of action given the state, $\pi(s) = p(a|s)$ initialized with uniform action probability.

The learning algorithm lets the BT run a fixed number of episodes (ξ), where each episode ends when the BT returns success or failure. Recall that failure can occur if the time limit is reached. During each episode, store the trace as a sequence of time-index state-action pairs $[(s(0), a(0)), \ldots, (s(m), a(m))]$ where m is the trace length. At the end of each episode, update the state action table using

$$p(a(t)|a(t)) \leftarrow \pi(a(t)|a(t)) + \mu^{m-t} * b, \tag{8}$$

where π is the policy, $(s(t), a(t))$ is the state-action pair at time t in the trace τ, $\mu = 0.9$ is the discount factor and b is a binary variable which is translated to +1 (-1) when $\mathbb{T}_\psi$ returns success (failure). After the update, each $p(a|s)$ is renormalized so that $\sum_a p(a|s) = 1$. Since every successful trace is guaranteed to satisfy Ψ^{C2H}, setting $b = 1$ makes actions observed in the trace more likely. It is not true that every failed trace does not satisfy Ψ^{C2H}, the setting $b = -1$ biases exploration to those policies that lead to success.

We can use the fact that Ψ^{C2H} is composed of two PPATasks connected by the **U** operator to use feedback from the PPATask subtree to learn different policies for ψ^{Cheese} and ψ^{Home}. Since the two tasks are sequential, the ψ^{Cheese} sub-tree needs to be successful for ψ^{Home} to be learned. Let $p(a|s, C)$ represent the phase where the agent is seeking the cheese, and let $p(a|s, H)$ represent the phase where the agent has the cheese and is learning to return home. The BT nodes $Action_{Cheese}$ and $Action_{Home}$ use $p(a|s, C)$ and $p(a|s, H)$, respectively. Divide the trace τ into two phases, $\tau = [\tau_C, \tau_H]$, where the first part of the trace attempts to perform the Cheese task and the second part attempts to perform the Home task.

Two conditions apply. First, if the cheese subtree never returns success, Eq. (8) updates $p(a|s, C)$ over the entire trace τ. Second, if the cheese subtree returns success, the τ_C is the part of the trace up to when the task is successful and is used to update $p(a|s, C)$. The remaining part of the trace is τ_H, which uses BT successes of failures to update $p(a|s, H)$.

Experiment Design and Results. All learning experiments use the following parameters, empirically selected to illustrate successful learning: start a location $s_{(3,0)}$, $m = 50, \xi = 200$, and $\mu = 0.9$ unless specified otherwise. Two dependent

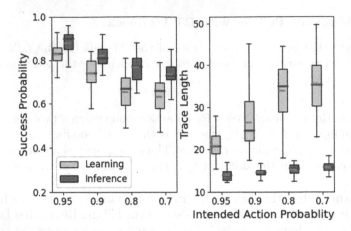

Fig. 6. Success probabilities and trace lengths for learning phase and inference phase.

measures are used: *learning success probability* and *inference success probability*. *Learning success probability* is defined as the average success status of mission BT while the policies are being learned divided by the number of episodes. Similarly, the *inference success probability* is the number of successful missions using learned policies, divided by the total number of simulation runs. The agent's starting location is randomized in the inference success evaluations, and 50 simulations were conducted using the learned policies.

For the learning experiments, 50 independent simulations were conducted. Figure 6 compares the performance of policies during learning and inference settings. The golden boxes in the left sub-plot of Fig. 6 show that learning $p(a|s)$ directly from the return status of the BT produces good policies for the sequential Ψ^{C2H} problem. Similarly, the golden boxes in the right sub-lot of Fig. 6 show that the *trace length* increases with a decrease in intended action probabilities. Figure 6 depicts two distinct properties: a) the success probabilities are higher for the inference phase than the learning phase, and b) the trace length increases much more rapidly during the learning phase than the inference as the intended action probability decreases. These properties are seen because, during the inference phase, just one episode is sufficient to test the policy, whereas the learning phase requires many episodes where the uncertainty in the intended action probabilities accumulates with each episode.

Comparing the results from Fig. 5 to Fig. 6 shows that using the direct feedback from the BT to create policies for the action nodes has higher success probabilities than using policy iteration when the rewards and goal do not align well.

7   Fetch Robot Example

This section demonstrates how a BT for an LTL_f-based achievement goal works on the *Fetch* robot. A Fetch robot is a mobile robot with a manipulator arm that has a) 7 degrees of freedom, b) a modular gripper with easy gripper swapping, c) a torso with adjustable height, and d) an ability to reach items on the floor. The camera is at the head of the robot, so during manipulation tasks, its arm blocks its field of view.

7.1 Problem Specification

The sequential problem for the robot is a variant of *key-door* [8] problem. A rectangular box of size 1ftx1ft is the *active region* where its vision system is actively scanning, and the area outside the box is a *passive region*. The task has three blocks with three different shapes: the red block is the key, the black block is the door, and the blue block is the prize. The goal for the robot is to a) locate the key and stack it on the top of the door block, b) move both key and door blocks together to the passive zone, and c) locate the prize and carry it to the passive zone. The robot perceives the world through its vision system and interacts with its arm.

An LTL_f formula for *key-door* (KD) mission is

$$\Psi^{KD} = \mathbf{U}(\ \mathbf{F}\ \psi^{Key})(\ \mathbf{U}\ \mathbf{F}\ \psi^{Door}\ \mathbf{F}\ \psi^{Prize})$$

where three PPA tasks have the structure from in Eq. 7

$$\psi^{Key} = \vee(\wedge NoErr\ KeyStacked)(\wedge(\wedge NoErr\ IsKeyDoor)$$
$$(\mathbf{U}\ VisibleKeyDoor\ \wedge\ Action_{KeyStacked}))$$
$$\psi^{Door} = \vee(\wedge NoErr\ KeyDoorPassive)(\wedge(\wedge NoErr\ KeyStacked)$$
$$(\mathbf{U}\ KeyStacked\ \wedge\ Action_{KeyDoorPassive}))$$
$$\psi^{Prize} = \vee(\wedge NoErr\ PrizePassive)(\wedge(\wedge NoErr\ PrizeVisible)$$
$$(\mathbf{U}\ KeyDoorPassive\ \wedge\ Action_{PrizePassive}))$$

where the *NoErr* proposition checks if the robot is throwing any system errors, *KeyStacked* proposition checks if the key and door block are stacked together, *IsKeyDoor* checks if the key and door block are on the table, *VisibleKeyDoor* checks if the key and door are visible and not overlapping, *KeyDoorPassive* checks if key and door are in passive area of the table, *PrizePassive* checks if the prize block is in passive area, and *PrizeVisible* checks if the prize block is in the active area. The state vector at time t is the vector of truth values for all the atomic proposition described above, $\mathbf{s}_t = [NoErr, KeyStacked, \ldots, PrizeVisible]$.

The action nodes $Action_{KeyStacked}$, $Action_{KeyDoorPassive}$, and $Action_{PrizePassive}$ execute the plans to stack the key on top of the block, move the stack of key and door to the passive area, and move the prize to passive area, respectively. The sensing and plans were created using widely available Fetch robot libraries [37].

7.2   Experiment Design

Two different modes of the experiment were conducted: *baseline* and *PPA-Task-LTLf* . For each mode, 25 independent robot trial was done. One trial corresponds to allowing the robot to perform the mission until it returns failure or success. In the *baseline* mode, the robot tried executing the mission without using LTL_f to BT decomposition, and the plans were connected using if-else code blocks. The second mode *PPA-Task-LTLf* used the Ψ^{KD} specification and the corresponding BT. For each mode, ten trials were completed without any external disturbances. For the other remaining 15 trials, a human physically interrupted the robot by removing or moving blocks. For each trial, the disturbance was only done once. The interruption was uniformly applied at each stage of the mission. Recall that the **F**inally operator from the mission grammar can send a reset signal to its child sub-tree. In all experiments below, if some tasks fail the robot can retry once, which was chosen subjectively as the robot generally accomplishes the mission after one try.

7.3   Results

The top part of Table 3 shows the performance of the robot for the key-door problem without using LTL_f to BT decomposition. The robot failed to complete the mission when interrupted in the baseline condition no matter when the interruption occurred. In the *baseline* mode experiments, the robot does not have the ability to resume from where it last failed because the baseline algorithm does not track details of its failures and attempt to correct constraint violations. Since the BT has a modular postcondition-precondition-action (PPA) structure, by design, it can resume from the previous failure point if allowed to retry.

The bottom part of Table 3 shows the robot's performance when mission BT is used. Despite human interruptions at different stages of the mission, most of the time, the robot could complete the mission as it was allowed to do one retry. The robot failed once on the ψ^{Door} task and twice on the ψ^{Prize} task after interruptions due to constraint violations that could not be reversed. The most important constraint violation was a violation of the *NoErr* proposition, which encodes robot system errors that arise when the planning sub-system is unable to generate plans for the current system states. The descriptive data in Table 3 suggest that the mission success rate is higher when the behavior tree implementation of the mission grammar was used, which was one of the reasons for implementing the mission grammar in a behavior tree.

Table 3. Experiment results under a) the baseline conditions and b) the BT obtained from Mission and Task grammars.

Baseline Conditions			
Status	Normal	Human Disturbances In	
		$Task^{\text{Key}}$	$Task^{\text{Door}}$	$Task^{\text{Prize}}$
Success	10	0	0	0
Failure	0	5	5	5
Behavior Tree Conditions			
Status	Normal	Human Disturbances In	
		$Task^{\text{Key}}$	$Task^{\text{Door}}$	$Task^{\text{Prize}}$
Success	10	5	4	3
Failure	0	0	1	2

8   Summary and Future Work

This paper presented a mission grammar designed for achievement-oriented goals that require temporal coordination among subtasks. The temporal constraints were formalized in the mission grammar, which produced linear-temporal logic formulas from a subset of LTL operators. The grammar for the tasks was constructed to use postcondition-precondition-action structures, allowing the construction of behavior trees that used these structures. Every successful trace produced by the behavior tree satifies the LTL goal.

A key structure of the behavior trees is that the action nodes can create plans using off-the-shelf planners, which is in contrast to many previous work on converting LTL formulas into state machines. The examples presented were straightforward demonstrations for how existing planners could be used to implement the action nodes. Some properties of the resulting planners were demonstrated, specifically the risk of reward-goal misalignment if using MDP-based planners, the ability to use the return status of the behavior tree to train state-action policies, and the ability to retry subtasks to produce more resilient behaviors. The most important piece of future work is to encode sophisticated goals for real robots performing complicated tasks, and then identify state-of-the-art planners that are most compatible with the type of feedback provided by the behavior tree.

Acknowledgements. This work was supported by the U.S. Office of Naval Research (N00014-18-1-2831). The authors thank Elijah Pettitt, who was an undergraduate research assistant, for programming and running the experiments with the Fetch robot.

References

1. Ahmadi, M., Sharan, R., Burdick, J.W.: Stochastic finite state control of POMDPs with LTL specifications. arXiv preprint arXiv:2001.07679 (2020)
2. Antoniotti, M., Mishra, B.: Discrete event models+ temporal logic= supervisory controller: automatic synthesis of locomotion controllers. In: Proceedings of 1995 IEEE International Conference on Robotics and Automation, vol. 2, pp. 1441–1446. IEEE (1995)
3. Bacchus, F., Kabanza, F.: Planning for temporally extended goals. Ann. Math. Artif. Intell. **22**(1–2), 5–27 (1998)
4. Barnat, J., et al.: How to distribute LTL model-checking using decomposition of negative claim automaton. In: SOFSEM, pp. 9–14 (2002)
5. Bertoli, P., Cimatti, A., Pistore, M., Roveri, M., Traverso, P.: MBP: a model based planner. In: Proceedings of the IJCAI-01 Workshop on Planning under Uncertainty and Incomplete Information (2001)
6. Biggar, O., Zamani, M.: A framework for formal verification of behavior trees with linear temporal logic. IEEE Robot. Autom. Lett. **5**(2), 2341–2348 (2020)
7. Biggar, O., Zamani, M., Shames, I.: On modularity in reactive control architectures, with an application to formal verification. ACM Trans. Cyber-Phys. Syst. (TCPS) **6**(2), 1–36 (2022)
8. Chevalier-Boisvert, M., Willems, L., Pal, S.: Minimalistic gridworld environment for gymnasium (2018). www.github.com/Farama-Foundation/Minigrid
9. Colledanchise, M., Murray, R.M., Ögren, P.: Synthesis of correct-by-construction behavior trees. In: 2017 IEEE/RSJ International Conference on Intelligent Robots and Systems (IROS), pp. 6039–6046. IEEE (2017)
10. Colledanchise, M., Ögren, P.: How behavior trees generalize the teleo-reactive paradigm and and-or-trees. In: 2016 IEEE/RSJ International Conference on Intelligent Robots and Systems (IROS), pp. 424–429. IEEE (2016)
11. Colledanchise, M., Ögren, P.: Behavior Trees in Robotics and AI: An Introduction. CRC Press, Boca Raton (2018)
12. Ding, X.C.D., Smith, S.L., Belta, C., Rus, D.: LTL control in uncertain environments with probabilistic satisfaction guarantees. IFAC Proc. Vol. **44**(1), 3515–3520 (2011)
13. Fainekos, G.E., Kress-Gazit, H., Pappas, G.J.: Hybrid controllers for path planning: a temporal logic approach. In: Proceedings of the 44th IEEE Conference on Decision and Control, pp. 4885–4890. IEEE (2005)
14. Fainekos, G.E., Kress-Gazit, H., Pappas, G.J.: Temporal logic motion planning for mobile robots. In: Proceedings of the 2005 IEEE International Conference on Robotics and Automation, pp. 2020–2025. IEEE (2005)
15. Favorito, M., Cipollone, R.: Flloat (2020). www.whitemech.github.io/flloat/
16. Jensen, R.M., Veloso, M.M.: OBDD-based universal planning for synchronized agents in non-deterministic domains. J. Artif. Intell. Res. **13**, 189–226 (2000)
17. Klein, J., Baier, C.: Experiments with deterministic ω-automata for formulas of linear temporal logic. Theoret. Comput. Sci. **363**(2), 182–195 (2006)
18. Lahijanian, M., Wasniewski, J., Andersson, S.B., Belta, C.: Motion planning and control from temporal logic specifications with probabilistic satisfaction guarantees. In: 2010 IEEE International Conference on Robotics and Automation, pp. 3227–3232. IEEE (2010)
19. Maretić, G.P., Dashti, M.T., Basin, D.: LTL is closed under topological closure. Inf. Process. Lett. **114**(8), 408–413 (2014)

20. Marzinotto, A., Colledanchise, M., Smith, C., Ögren, P.: Towards a unified behavior trees framework for robot control. In: 2014 IEEE International Conference on Robotics and Automation (ICRA), pp. 5420–5427. IEEE (2014)
21. Parr, R., Russell, S.J.: Reinforcement learning with hierarchies of machines. In: Advances in Neural Information Processing Systems, pp. 1043–1049 (1998)
22. Piterman, N., Pnueli, A., Sa'ar, Y.: Synthesis of reactive (1) designs. In: Emerson, E.A., Namjoshi, K.S. (eds.) VMCAI 2006. LNCS, vol. 3855, pp. 364–380. Springer, Heidelberg (2006). https://doi.org/10.1007/11609773_24
23. Pnueli, A.: The temporal logic of programs. In: 18th Annual Symposium on Foundations of Computer Science (SFCS 1977), pp. 46–57. IEEE (1977)
24. Pnueli, A., Rosner, R.: On the synthesis of a reactive module. In: Proceedings of the 16th ACM SIGPLAN-SIGACT Symposium on Principles of Programming Languages, pp. 179–190 (1989)
25. Rozier, K.Y., Vardi, M.Y.: A multi-encoding approach for LTL symbolic satisfiability checking. In: Butler, M., Schulte, W. (eds.) FM 2011. LNCS, vol. 6664, pp. 417–431. Springer, Heidelberg (2011). https://doi.org/10.1007/978-3-642-21437-0_31
26. Russell, S.J.: Artificial Intelligence a Modern Approach. Pearson Education Inc., London (2010)
27. Sadigh, D., Kim, E.S., Coogan, S., Sastry, S.S., Seshia, S.A.: A learning based approach to control synthesis of Markov decision processes for linear temporal logic specifications. In: 53rd IEEE Conference on Decision and Control, pp. 1091–1096. IEEE (2014)
28. Schillinger, P., Bürger, M., Dimarogonas, D.V.: Decomposition of finite LTL specifications for efficient multi-agent planning. In: Grob, R., et al. (eds.) Distributed Autonomous Robotic Systems, pp. 253–267. Springer, Cham (2018). https://doi.org/10.1007/978-3-319-73008-0_18
29. Sistla, A.P.: Safety, liveness and fairness in temporal logic. Formal Aspects Comput. **6**(5), 495–511 (1994)
30. Stonier, D., Staniasnek, M.: Py-trees (2020). www.py-trees.readthedocs.io/en/devel/index.html
31. Sutton, R.S., Precup, D., Singh, S.: Between MDPs and semi-MDPs: a framework for temporal abstraction in reinforcement learning. Artif. Intell. **112**(1–2), 181–211 (1999)
32. Tadewos, T.G., Newaz, A.A.R., Karimoddini, A.: Specification-guided behavior tree synthesis and execution for coordination of autonomous systems. Expert Syst. Appl. **201**, 117022 (2022)
33. Toro Icarte, R., Klassen, T.Q., Valenzano, R., McIlraith, S.A.: Teaching multiple tasks to an RL agent using LTL. In: Proceedings of the 17th International Conference on Autonomous Agents and MultiAgent Systems, pp. 452–461 (2018)
34. Van Riemsdijk, M.B., Dastani, M., Winikoff, M.: Goals in agent systems: a unifying framework. In: Proceedings of the 7th International Joint Conference on Autonomous Agents and Multiagent Systems-Volume 2, pp. 713–720. International Foundation for Autonomous Agents and Multiagent Systems (2008)
35. Vasile, C.I., Belta, C.: Sampling-based temporal logic path planning. In: 2013 IEEE/RSJ International Conference on Intelligent Robots and Systems, pp. 4817–4822. IEEE (2013)

36. Vazquez-Chanlatte, M., Jha, S., Tiwari, A., Ho, M.K., Seshia, S.: Learning task specifications from demonstrations. In: Advances in Neural Information Processing Systems, pp. 5367–5377 (2018)
37. Wise, M., Ferguson, M., King, D., Diehr, E., Dymesich, D.: Fetch and freight: standard platforms for service robot applications. In: Workshop on Autonomous Mobile Service Robots, pp. 1–6 (2016)

Discovery and Analysis of Rare High-Impact Failure Modes Using Adversarial RL-Informed Sampling

Rory Lipkis[✉] and Adrian Agogino

Intelligent Systems Division, NASA Ames Research Center, Moffett Field, CA, USA
{rory.lipkis,adrian.k.agogino}@nasa.gov

Abstract. Adaptive learning agents have tremendous potential to handle critical tasks currently performed by humans. Unfortunately, due to their complexity, it can be difficult to verify that these learning agents do not have critical failure modes. Standard verification and validation methods often do not apply directly to learning agents and Monte Carlo methods have difficulty covering even a small fraction of the state space, especially in multiagent systems or over long time horizons. To overcome this difficulty, we demonstrate an adaptive stress-testing method based on reinforcement learning of correlations that raise the probability of failure. This approach has three key properties: (1) it is able to find rare failure modes with far greater sample efficiency than Monte Carlo methods, (2) it can estimate the true probability of a failure mode despite the inherent bias in the learning method, and (3) it is capable of learning and resampling compact representations of multimodal failure spaces. These properties are important in practice as we need to find disparate failure modes while accounting for their actual relevance. This is a significant advantage over traditional adaptive stress testing methods that give abstract likelihoods of particular failure instances, but cannot estimate the probability of a broader failure mode. We test our algorithm on a simple problem from the aviation domain where an autonomous aircraft lands in gusty wind conditions. The results suggest that we can find failure modes with far fewer samples than the Monte Carlo approach and simultaneously estimate the probability of failure.

Keywords: Reinforcement learning · validation · sampling · statistical failure analysis

1 Introduction

Validation of complex stochastic systems is challenging. Methods such as Monte Carlo sampling often fail, since for large complex systems, a very small fraction of possible outcomes can be realistically sampled. Indeed, even learning problems involving simple sequences of actions can have enormous state spaces that compound with each additional step. The problem becomes even worse in the case of multiagent systems.

© The Author(s), under exclusive license to Springer Nature Switzerland AG 2024
F. Amigoni and A. Sinha (Eds.): AAMAS 2023 Workshops, LNAI 14456, pp. 123–140, 2024.
https://doi.org/10.1007/978-3-031-56255-6_7

To address these issues, we propose to use a stochastic validation process based on reinforcement learning. Validation of a system under test (SUT) entails determining through testing and analysis whether specified requirements are met. Traditionally, statistical validation (falsification) is performed by randomly sampling system inputs and transitions, producing an estimate of the failure likelihood. This Monte Carlo approach becomes computationally infeasible when validating the performance of complex systems over longer horizons. For example, in the case of collision avoidance systems, a common target of verification efforts, system failure might only result from the unlikely confluence of reckless operation and bad luck over an extended period of time.

Detecting such complicated failure modes is a difficult task. Since direct sampling explores regions in proportion to their likelihood of occurrence, the majority of computational effort is expended evaluating near-nominal system behavior. Any amount of model error – neglecting a small interstep correlation, for instance – can render a valid failure mode undiscoverable. In the event that a failure event is detected, it may not recur sufficiently to characterize the larger mode.

To mitigate this problem, it is common to manually bias the search in a manner that takes advantage of domain knowledge to expedite failure discovery. Though successful in eliciting higher numbers of failures, this technique arguably jeopardizes the notion of *independence* in verification and validation. If a prejudice towards expected failure modes is built into the testing methodology, the tester potentially forecloses on discovering unknown failure modes. For safety-critical systems, these may represent the failures of most concern.

Adversarial testing provides a compelling solution to this dilemma. As AI systems become more widespread, much research has focused on generating adversarial attacks against deep neural networks in decision and perception subsystems. For instance, it has long been observed that classifier accuracy can be vulnerable to small perturbations in the input, particularly when a model has not been trained for robustness [3].

In recent years, many efforts have applied the idea of adversarial testing to autonomous control systems in simulation. Intelligent test case generation has been used to find static environment parameters that challenge self-driving vehicle algorithms [18,20]. In such cases, different combinations of weather conditions, sensor faults, and pedestrians are explored until critical requirements are violated. A more fine-grained testing strategy involves step-wise perturbations. In particular, a controller that makes decisions based on its surroundings can be pitted against an adversarial agent in the same environment whose goal is to force violations [2,19]. In such experiments, the opponent's disruptive ability must be limited if the goal is to produce realistic failure cases, often achieved with a handpicked heuristic.

In adaptive stress testing (AST), the adversarial agent is the environment itself, which is taken to be a probabilistic model of perturbations [12]. This induces a natural measure of likelihood that is factored into the agent's reward function to limit its adversarial capacity. AST thus provides a framework for learning the *likeliest* failures of an autonomous SUT with reinforcement learning. Information about unsafe regions of the state space can be automatically gathered and exploited, allowing failure modes to be discovered more efficiently while maintaining a degree of objectivity. This approach has proven useful for risk characterization and system development, and has seen use in a variety of applications, including aircraft collision avoidance [12, 13], autonomous driving [1, 7], trajectory planning for small unmanned aircraft [11], autonomous aircraft taxiing [5], and flight management [15].

In this paper, we present a framework that combines the likelihood awareness of AST with explicit policy learning and importance resampling, providing several advantages: (1) rare and disparate failure modes can be efficiently discovered, (2) failure mode probabilities can be estimated with low bias, and (3) learned failure modes can be resampled, generating random failure cases with minimal additional computation. These benefits allow for a much more comprehensive analysis of failure events in complex systems.

2   Adaptive Stress Testing

In the AST formulation, the SUT interacts with a stochastic environment within a simulation. The simulation is summarized by a state $s \in \mathcal{S}$; failure corresponds to some subset $\mathcal{F} \subset \mathcal{S}$. The environment consists of a set of external disturbances collected into the random variable $X \in \mathcal{X}$, where $X \sim p(x)$. It is helpful to specify some sort of distance-to-failure metric $d(s)$ to guide the learning. Recent research has demonstrated success in the absence of such a heuristic, although this requires a significantly more specialized solution technique [8].

As an example, in the aircraft collision avoidance setting, the state might contain the positions and velocities of several aircraft while the environment describes externalities (from the perspective of the SUT) such as pilot controls, wind gusts, or sensor noise. A sensible distance metric would be the minimum pairwise distance between aircraft.

Rather than sampling values from the environment as in traditional sample-based testing, AST explicitly optimizes over the disturbances to find the most likely failure events in the SUT. For an arbitrary T-step trajectory, the joint likelihood is given by

$$p(s_0, \ldots, s_T) = p(s_0) \prod_{t=1}^{T} p(s_t \mid s_0, \ldots, s_{t-1})$$

$$= p(s_0) \prod_{t=1}^{T} p(s_t \mid s_{t-1})$$

$$= p(s_0) \prod_{t=0}^{T-1} p(x_t \mid s_t), \tag{1}$$

where simplifications are due to the Markov property and the assumption that all randomness is captured in the specification of the environment (i.e., the SUT is either deterministic or derandomizable). Thus, the high-level goal of AST is to solve the optimization problem

$$\max_{x_0,\ldots,x_{T-1}} \prod_{t=0}^{T-1} p(x_t \mid s_t)$$

$$\text{subject to } s_T \in \mathcal{F} \tag{2}$$

for a fixed initialization s_0. By considering failure states to be absorbing, one may account for failures that occur at any $t \leq T$. Note that this problem amounts to the maximization of a relatively simple objective function subject to an arbitrarily complex constraint. The likely non-convexity (and even disconnectedness) of the feasible region suggests that classical optimization methods may be insufficient.

In AST, the problem is formulated quite naturally as a Markov decision processes (MDP), enabling the use of reinforcement learning techniques. Figure 1 illustrates the typical AST architecture. At each time step, the agent observes the state of the simulation, selects an action (an environment instance), and receives a reward

$$r(s,x) = \begin{cases} r_f & \text{if } s \in \mathcal{F} \\ -d_{\min} & \text{if } s \notin \mathcal{F}, \ t = T \\ \log p(x \mid s) & \text{if } s \notin \mathcal{F}, \ t < T, \end{cases} \tag{3}$$

where $d_{\min}$ is the closest distance to failure achieved across the entire sequence of states in the reinforcement learning episode. This formulation encourages solutions to prioritize likelier failures, and amounts to a softened version of the original problem, since the constraint is replaced by a penalty.

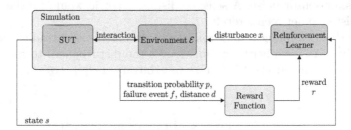

Fig. 1. Adaptive stress testing architecture. A generic reinforcement learning agent chooses instances of a stochastic environment to elicit the likeliest possible failure in the system under test. The top-scoring failure traces are returned.

In the original description of AST, the MDP is solved with the Monte Carlo tree search (MCTS) algorithm, which builds a tree of actions and recursively updates its value estimates as it explores different paths [6]. This algorithm has the advantage of anonymizing states: if an explicit representation of the environment is inaccessible but the random number generator of the simulation can be seeded, learning can still be performed over the space of seed sequences; the state s_t is effectively the history of seeds used up to step t.

2.1   Limitations

AST has historically achieved excellent results, as is clear from its use by system designers and researchers across governmental, industrial, and academic domains. However, the original framework has several important limitations, in part due to its output consisting of a set of failure traces (sequences of environment values that lead to SUT failure).

Modality. AST can find failures much more quickly than a basic Monte Carlo search. However, these failures are typically very similar and ultimately converge on the mode (or a local maximum) of the conditional probability distribution

$$p_{\mathcal{F}}(x) = p(x_0, \ldots, x_{T-1} \mid s_T \in \mathcal{F}). \tag{4}$$

The diversity of the output depends on the extent to which the particular learning algorithm explores while seeking the global optimum, as well as its capacity for storing suboptimal solutions; it is not a deliberate feature of the search. This is a result of the formulation of the MDP and is asymptotically independent of the specific solution method: with enough training, the top k solutions will all be minor variations of the optimum.

Generality. The inherent unimodality of the framework motivates a paradigm of running AST multiple times in parallel from different initializations with the hope of landing in a different failure mode – in essence, conducting a Monte Carlo search of the initialization space. Since MCTS builds failures sequentially from an initial root, its output does not apply at any other starting point. This results in a large amount of computational waste, since information about the search space is not shared between processes. As a result, similar solutions cannot be learned in a manner that takes advantage of that similarity. Automated categorization techniques have shown potential in structuring a set of failure traces based on feature similarity [10]. While useful for facilitating human interpretation of AST results, this is a purely extrinsic approach that does not refer to the underlying model. Recent research has explored refining low-fidelity AST results with a backwards retraining scheme [9]. This approach represents a form of generalization, but its focus is on perfecting existing failure traces.

Statistics. AST cannot generate valid failure probabilities. For any given failure, the corresponding joint probability density value is calculable, but this is not a meaningful quantity in either absolute or relative terms – worse, such values can be easily misinterpreted as indicating exceptionally low risk for a system, especially when reported without context. Because a point evaluation of the probability density function does not reflect the shape of the surrounding mode and the failures do not have a strongly relational representation, the desire to account for probability mass from arbitrarily similar failures is unrealizable.

3   Failure Trace Resampling

Calculating failure statistics from conventional Monte Carlo sampling is straightforward, as the failure probability is estimated by the ratio of failures to the total number of samples. However, this approach cannot be used directly with AST, which generates failures in a deliberately nonuniform process. This can be rectified by using its output as the basis of an importance sampling scheme, which involves drawing from an alternative distribution and reweighting samples to correct for their missized contribution to the estimate. This approach makes it possible to calculate the likelihood of an entire failure mode along with a confidence interval.

Consider the AST trace $x^* = [x_1^*, x_2^*, \ldots x_T^*] \in \mathcal{X}^T$, a high-likelihood environment sequence resulting in SUT failure[1]. Since the failure region may be arbitrary complex, it is defined implicitly by an indicator function $\mathbf{1}_{\mathcal{F}}(s) \in \{0, 1\}$. For a fixed initialization and environment sequence x, one can alternatively consider the indicator function $f(x) = \mathbf{1}_{\mathcal{F}}(s(x))$, where $s(x)$ is the state after the application of x. Let $X = [X_1, X_2, \ldots X_T]$ be the random trace corresponding to a T-step "rollout" of the environment. Then, $f(X)$ is a binary random variable indicating whether or not a failure occurs during the trace. The overall failure prevalence is given by

$$\mu = P(f(X) = 1) = \mathbf{E}[f(X)] = \int_{\mathcal{X}^T} f(x)p(x)\,dx\,, \tag{5}$$

where $p(x) = \prod_{i=1}^{T} p(x_i)$ denotes the joint probability distribution of the trace and $dx = dx_1 \wedge \cdots \wedge dx_T$. In theory, this integral could be approximated via Monte Carlo integration, in which

$$\mathbf{E}[f(X)] \approx \frac{1}{n} \sum_{i=1}^{n} f(x^{(i)})\,, \tag{6}$$

where sample traces $x^{(i)}$ are drawn from X. However, when failures are sufficiently rare, i.e., $\mu \ll 1/n$, the estimate may be highly inaccurate or simply zero.

Instead, let $X^* \sim q(x)$ be a surrogate random variable that prioritizes x^* in some way and satisfies the coverage condition that $q(x) > 0$ wherever $p(x) > 0$ and $f(x) = 1$. The probability of failure can then be rewritten as

$$\mu = \int_{\mathcal{X}^T} f(x)\frac{p(x)}{q(x)}q(x)\,dx = \mathbf{E}\left[f(X^*)\frac{p(X^*)}{q(X^*)}\right]\,. \tag{7}$$

---

[1] The time index has been shifted to simplify the notation.

This expectation is performed over a distribution for which failures are by construction less rare; it is realized by the estimator

$$\hat{\mu} = \frac{1}{n} \sum_{i=1}^{n} f(x^{(i)}) \frac{p(x^{(i)})}{q(x^{(i)})} , \tag{8}$$

where sample traces $x^{(i)}$ are now drawn from X^*, i.e., rolled out from a surrogate environment. This quantity represents the probability of the failure mode in which x^* lies.

Surrogate Construction

The success of the transformation depends heavily on the choice of surrogate. A desirable surrogate distribution emphasizes known failures without sacrificing the variance of the original distribution. The theoretically optimal surrogate $q^*(x)$ is exactly proportional to $f(x)p(x)$; this distribution minimizes the resulting estimation variance but is difficult to obtain in a usable form [17].

However, using Bayes' theorem, we can expand the conditional probability distribution maximized by AST as

$$\begin{aligned} p_{\mathcal{F}}(x) &= P(X = x \mid f(X) = 1) \\ &= P(f(X) = 1 \mid X = x) \frac{P(X = x)}{P(f(X) = 1)} \\ &= f(x)p(x)/\mu . \end{aligned} \tag{9}$$

Consequently, AST output converges to the statistical mode of a distribution that is equal to $q^*(x)$ up to a constant of proportionality. We can therefore use this output as the basis of an approximation of the optimal surrogate.

Since $p(x)$ may be arbitrarily complex or entirely blackboxed, there are limited options for $q(x)$ to be constructed generally and systematically. One reasonable method is to create a mixture model between the original distribution and a shifted variant. If X is continuous and unbounded, let $X^* = X - \mathbf{E}[X] + x^*$ be the random variable with elementwise probability distribution

$$q_{\mathrm{sh}}(x_t) = p\left(x_t + \mathbf{E}[X_t] - x_t^*\right) . \tag{10}$$

This is a version of the original distribution shifted to center around the high-likelihood AST trace. It has the benefit of not requiring any knowledge of the functional form of $p(x)$, other than its expectation, which may be omitted if necessary (or if known to be zero). To sample X^*, one needs only to sample X and add the appropriate offset. If X is bounded in some way, the shifting approach will not work and a handpicked distribution such as a truncated normal may be required to satisfy the coverage condition. Note that $q_{\mathrm{sh}}(x)$ cannot approximate the shape of $q^*(x)$, only match its mode.

The approach can be augmented by using a mixture model that samples from either $p(x)$ or $q_{\mathrm{sh}}(x)$ at each step of the rollout, resulting in a distribution

$$q(x_t) = \epsilon p(x_t) + (1 - \epsilon)p\left(x_t + \mathbf{E}[X_t] - x_t^*\right) . \tag{11}$$

This modification limits the estimation variance by bounding the ratio in the integrand. The full scheme is described in Algorithm 1, where $\mathbf{E}[X]$ is assumed to be zero.

Algorithm 1. Importance resampling of a failure trace

Input: s_0, x^*, ϵ, T, N
Output: $\hat{\mu}$
 1: $\Sigma \leftarrow 0$ ▷ Accumulator
 2: **for** $i = 1$ to N **do**
 3:     $w \leftarrow 1$
 4:     $s_t \leftarrow s_0$ ▷ Fixed initialization
 5:     **for** $t = 1$ to T **do**   ▷ Fixed horizon
 6:         $x_t \leftarrow$ **Sample**$[p(x)]$
 7:         $\alpha \leftarrow$ **Sample**$[\text{Unif}(0, 1)]$
 8:         **if** $\epsilon < \alpha$ **then**   ▷ Mixture model
 9:             $x_t \leftarrow x_t + x_t^*$
10:         **end if**
11:         $w \leftarrow wp(x_t)/(\epsilon p(x_t) + (1 - \epsilon)p(x_t - x_t^*))$
12:         **if not** **IsFailure**(s_t) **then**
13:             $s_t \leftarrow$ **Step**(s_t, x_t)
14:         **end if**
15:     **end for**
16:     **if** **IsFailure**(s_t) **then**
17:         $\Sigma \leftarrow \Sigma + w$
18:     **end if**
19: **end for**
20: **return** Σ/N ▷ Sample mean

3.1   Error Estimation

Naturally, estimation variance causes the estimate to differ from the true failure probability. This effect can be greatly exacerbated by the importance sampling scheme, which involves a potentially divergent ratio of probability densities. Nonetheless, error can be captured by probabilistic bounds.

Note that the estimator is theoretically unbiased, since its expectation is equal to the true probability. If $\hat{\mu}$ is the random variable corresponding to the estimated probability, then

$$\mathbf{E}[\hat{\mu}] = \mathbf{E}\left[\frac{1}{n}\sum_{i=1}^{n} f(X^{(i)})\frac{p(X^{(i)})}{q(X^{(i)})}\right]$$
$$= \mathbf{E}\left[f(X^*)\frac{p(X^*)}{q(X^*)}\right] = \mu, \tag{12}$$

using the linearity of expectations and the fact that $X^{(i)} \sim X^*$ for all i. A similar analysis yields the variance of the estimation

$$\mathrm{Var}[\hat{\mu}] = \mathrm{Var}\left[\frac{1}{n}\sum_{i=1}^{n} f(X^{(i)})\frac{p(X^{(i)})}{q(X^{(i)})}\right]$$
$$= \frac{1}{n}\mathrm{Var}\left[f(X^*)\frac{p(X^*)}{q(X^*)}\right], \tag{13}$$

using the scaling and linearity properties of variance. This value can itself be estimated as

$$\widehat{\sigma^2} = \frac{1}{n(n-1)}\sum_{i=1}^{n}\left[f(x^{(i)})\frac{p(x^{(i)})}{q(x^{(i)})} - \hat{\mu}\right]^2, \tag{14}$$

where the $n-1$ term is the standard bias correction. As the number of samples drawn from X^* increases, the distribution of the estimator $\hat{\mu}$ slowly approaches a normal distribution centered on the true failure probability μ with variance given by the above expression [17]. This yields the standard concentration bound

$$P(|\mu - \hat{\mu}| \geq \delta) \approx 2\left(1 - \Phi\left(\frac{\delta}{\sqrt{\widehat{\sigma^2}}}\right)\right), \tag{15}$$

equivalently expressed as the confidence interval

$$P\left(|\mu - \hat{\mu}| \geq \sqrt{\widehat{\sigma^2}}\,\Phi^{-1}\left(1 - \frac{\epsilon}{2}\right)\right) \approx \epsilon. \tag{16}$$

Since the importance sampling scheme may overly concentrate within a mode of failure, the estimate tends to underestimate the true probability. It may occasionally be useful to independently upper-bound the overall failure probability with a standard Monte Carlo computation, even if it fails to yield a single failure. Applying a Chernoff bound to the sampling process yields

$$P(\mu - \hat{\mu}_{\mathrm{mc}} \geq \delta) \leq e^{-n\delta^2/2}. \tag{17}$$

Alternatively, an exact upper bound can be derived from Bayesian principles. If the sampling yields $\hat{\mu}_{\mathrm{mc}} = n/k$, then

$$P(\mu - \hat{\mu}_{\mathrm{mc}} \geq \delta) = I_{1-\hat{\mu}_{\mathrm{mc}}-\delta}\left(n - k + \tfrac{1}{2}, k + \tfrac{1}{2}\right), \tag{18}$$

where $I_x(a, b)$ is the regularized incomplete beta function. These bounds can be factored into the calculation of a refined confidence interval.

3.2 Multimodal Failure Trace Resampling

Multiple failure traces generated by AST can be combined to form a composite surrogate distribution. For a set of failure trajectories $\{x_1^*, x_2^*, \ldots x_k^*\}$, we can define

$$q(x) = \frac{1}{k} \sum_{i=1}^{k} q_i(x) , \tag{19}$$

where the $q_i(x)$ are the corresponding surrogate distributions. An issue arises when some failure traces are shorter than others. For the interpretation to remain valid, shorter traces can be padded randomly at sample time.

The distribution $q(x)$ should not be sampled directly, since this would nullify the correlated nature of each failure mode. For example, distributions centered about failure trajectories $x_+^* = 1$ and $x_-^* = -1$ must be sampled separately to produce trajectories that reach either failure mode; mixing them produces a zero-centered distribution.

To rectify this issue, $q(x)$ must be sampled as a mixture model: a distribution q_i is selected uniformly at random, then sampled. The full probability $q(x)$ is still used in the expectation estimate, as

$$\hat{\mu} = \frac{k}{n} \sum_{i=1}^{n} f(x^{(i)}) \frac{p(x^{(i)})}{\sum_{j=1}^{k} q_j(x^{(i)})} . \tag{20}$$

4   Failure Policies

In the previous sections, it was assumed that all rollouts begin from the same initial state. This restriction allows a tree-based reinforcement learning algorithm such as MCTS to efficiently find paths to failure. However, if the system is sufficiently transparent, it is possible to formulate a much more powerful approach to the AST problem. Instead of finding a set of paths to failure, we learn an optimally adversarial policy π^* that maps a state to the likeliest environment value that induces a path to failure. Instead of an initial state s_0, we specify an initial distribution $p_0(s)$ with support $\mathcal{S}_0 \subseteq \mathcal{S}$, which is sampled at the start of each training episode.

To accommodate a more continuous setting, the AST reward function is modified as

$$r(s, x, s') = \log p(x \mid s) + \Delta(s, s') + r_f \cdot \mathbf{1}_{\mathcal{F}}(s) , \tag{21}$$

where r_f is a bonus for reaching failure and

$$\Delta(s, s') \propto d(s) - d(s') \tag{22}$$

is a reward shaping term to guide the learning agent more efficiently towards failure. Since the term represents a conservative potential, i.e., the gradient of a scalar function of the MDP state, its addition to the reward function is policy-invariant [16]. A wide variety of reinforcement learning algorithms can be used to solve this MDP. Deep reinforcement learning offers an attractive option when state and environment spaces are high-dimensional. Due to the ability of neural networks to interpolate and generalize, this approach allows failure paths to be approximated between samples.

Proposition 1. *With sufficient training, the policy π^* is weakly guaranteed to capture all failure modes $\mathcal{F}_k$ for which $\mathcal{F}_k \cap \mathcal{S}_0$ is non-empty.*

Proof. Since the AST framework is only concerned with the likeliest failures of a system, this bias is reflected in the failure policy. As long as a failure region $\mathcal{F}_k$ intersects with $\mathcal{S}_0$, there must exist a region of initialization $\mathcal{F}_k \subseteq \mathcal{S}_k \subseteq \mathcal{S}_0$ from which it is the likeliest failure. Since all such regions will be sampled as $n \to \infty$ and the optimal policy π^* by construction produces the most likely path to failure, then π^* must represent all F_k, provided its underlying representation has sufficient expressive capacity[2]. The universal approximation theorem establishes that any degree of expressiveness can be achieved by a neural network of sufficient width and depth; similar guarantees can be formulated for a table-based policy. Furthermore, asymptotic convergence to the optimal policy is provable for certain well-known algorithms [4,21].

In practice, the policy representation and the training set are of finite size, so optimality arguments are largely theoretical. However, they underscore the fact that multiple independent failure modes can be discovered and latently represented in a failure policy. This opens the door to analyzing the learned policy, which can be used for its generative properties: randomly sampling the initialization space and rolling out π^* efficiently produces a stream of unique failure traces.

5 Failure Policy Resampling

The importance sampling scheme developed for traces can be extended naturally to failure policies. At each step, instead of considering perturbations around an optimal environment x_t^*, we now consider perturbations around $\pi^*(s_t)$, the output of the optimal policy. To formalize this difference, we consider a surrogate random policy Π^*. This may be accomplished in same manner as before, through shifting or manual selection. In the case of a shifted surrogate, the policy evaluation $\Pi^*(s)$ represents a distribution over environments that behaves according to the mixture model

$$q_s(x_t) = \epsilon p(x_t) + (1 - \epsilon)p\left(x_t + \mathbf{E}[X_t] - \pi^*(s)\right) . \qquad (23)$$

Then, for a given initial state s_0 and horizon T, the resulting rollout is represented by the length-T random vector $X_{s_0}^*$ with components

$$X_{s_0,t}^* = \Pi^*(s_t), \qquad (24)$$

---

[2] It should be noted that even an optimal policy cannot necessarily capture all possible failures. Each initialization is associated with a single failure region: if from point s_0 the system admits failures $\mathcal{F}_1$ and $\mathcal{F}_2$, the policy will capture only the likelier outcome.

and the corresponding joint probability is

$$q(x_{s_0}^*) = \prod_{t=1}^{T} q(x_{s_0,t}^*). \tag{25}$$

The expected probability of failure from an initial state is thus

$$\mu(s_0) = \mathbf{E}_x\left[f(X_{s_0}^*)\frac{p(X_{s_0}^*)}{q(X_{s_0}^*)}\right]. \tag{26}$$

Finally, the overall failure probability can be written as

$$\mu = \mathbf{E}\left[\mu(S_0)\right] = \mathbf{E}_{s_0,x}\left[f(X_{S_0}^*)\frac{p(X_{S_0}^*)}{q(X_{S_0}^*)}\right], \tag{27}$$

where the subscripts indicate a joint expectation over initialization and environment spaces, respectively. The expectation is realized with the estimate

$$\hat{\mu} = \frac{1}{n}\sum_{i=1}^{n} f\left(x_{s_0}^{(i)}\right)\frac{p\left(x_{s_0}^{(i)}\right)}{q\left(x_{s_0}^{(i)}\right)}, \tag{28}$$

where each initial state s_0 is drawn from S_0 and the subsequent rollout x_{s_0} is drawn from $X_{s_0}^*$, applying the random policy step by step. The principle of deferred decisions ensures that this sequential sampling is equivalent to randomly sampling the entire trace at once [14]. As in the previous section, it is important not to terminate rollouts if an error occurs prematurely; the rollout should be padded to the horizon with actions sampled from $X_{s_0}^*$. The full algorithm is described in Algorithm 2, again assuming zero-mean environment variables.

The variance analysis is unchanged in the policy setting, so the confidence interval remains valid; however, the potential for high sample variance grows significantly. For an general Monte Carlo computation, the standard deviation of the estimate is equal to $\sigma/\sqrt{n}$, where σ is the true standard deviation of the quantity of interest. Since we are now considering a product sample space, σ is greatly increased and it is important to increase the sample size n accordingly. The estimate could also be performed with a double loop, making explicit the need to cover both spaces sufficiently. Luckily, as with most Monte Carlo schemes, the computation lends itself naturally to multiprocessing optimizations.

Algorithm 2. Importance resampling of a failure policy

Input: S_0, π^*, ϵ, T, N
Output: $\hat{\mu}$
1: $\Sigma \leftarrow 0$ ▷ Accumulator
2: **for** $i = 1$ to N **do**
3:     $w \leftarrow 1$
4:     $s_t \leftarrow$ **Sample**$[S_0]$ ▷ Random initialization
5:     **for** $t = 1$ to T **do**   ▷ Fixed horizon
6:        $x_t \leftarrow$ **Sample**$[p(x)]$
7:        $x_t^* \leftarrow \pi^*(s_t)$ ▷ Policy query
8:        $\alpha \leftarrow$ **Sample**$[\text{Unif}(0,1)]$
9:        **if** $\epsilon < \alpha$ **then**   ▷ Mixture model
10:          $x_t \leftarrow x_t + x_t^*$
11:        **end if**
12:        $w \leftarrow wp(x_t)/(\epsilon p(x_t) + (1 - \epsilon)p(x_t - x_t^*))$
13:        **if not IsFailure**(s_t) **then**
14:          $s_t \leftarrow$ **Step**(s_t, x_t)
15:        **end if**
16:     **end for**
17:     **if IsFailure**(s_t) **then**
18:        $\Sigma \leftarrow \Sigma + w$
19:     **end if**
20: **end for**
21: **return** Σ/N ▷ Sample mean

---

6 Experimental Results

To demonstrate the fundamental abilities of the extended framework, we consider a simple toy problem in which a small aircraft autonomously lands on a runway in gusty conditions. Episodes are initialized with the aircraft approaching the runway and deviating slightly from the center-line: $x_1 = 0$ and $x_2 \sim \mathcal{N}(0, \sigma_i^2)$. Episodes terminate when the aircraft reaches the start of the runway ($x_1 > a$). The environment consists of stochastic cross-track transitions $\Delta x_2 \sim \mathcal{N}(0, \sigma_t^2)$, and failure occurs if the aircraft lands too far from the center-line, a region defined as

$$\mathcal{F} = \{x_1, x_2 \in \mathbf{R} \mid x_1 > a, |x_2| > b\}. \tag{29}$$

The system is visualized in Fig. 2, where failure zones are shown in hatched red and non-failure terminating zones in green. Though the behavior of the SUT is completely trivial (it takes no actions to stabilize the trajectory), the toy problem is useful as it exhibits two distinct modes of failure with closed-form likelihoods. This allows the basic correctness of the framework to be tested against an analytical result: for the system shown in Fig. 2, the overall probability of failure is

$$\mu = 2\left(1 - \Phi\left(\frac{b}{\sqrt{\sigma_i^2 + \lceil a/v \rceil \sigma_t^2}}\right)\right) \approx 4.023 \times 10^{-13}. \tag{30}$$

The low probability rules out a direct Monte Carlo approach, since each failure would on average require simulating over 10^{12} episodes; a meaningful estimate of the probability would require at least another order of magnitude.

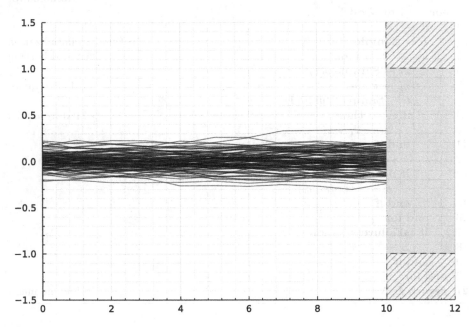

Fig. 2. 100 random trajectories of the example system, with zero recorded failures. Parameters are $a = 10$, $b = 1$, $\sigma_i = 0.1$, $\sigma_t = 0.03$, and $v = 1$; the effective horizon is 10. Although the effect of the stochastic environment is visible, a sustained disturbance would be required to elicit failure.

We solve the MDP with the soft actor-critic algorithm[3], yielding the policy shown in Fig. 3. The policy and value networks both contain two hidden layers of size 100. Training was performed for 2.5×10^4 episodes with a learning rate of 10^{-4}. The failure policy was then passed into the analysis stage, which resampled 10^7 additional episodes with an $\epsilon = 0.05$ mixture model. The estimated failure probability is approximately 3.253×10^{-13}, 19% less than the true probability, with an estimate standard deviation of 1.378×10^{-13}.

It should be noted that this standard deviation is quite high, given the scale of the estimate. The $\pm 3\sigma$ neighborhood encompasses all values from 0 to 7.386×10^{-13}, implying that there is a nearly 50% chance that the failure probability is arbitrarily low. This is a consequence of the fundamentally reciprocal nature of probability, which is not accounted for by this form of estimation but is crucial to how probabilities are used and understood. This issue is described further in

---

[3] Experiments with the simpler Q-learning algorithm yielded a nearly identical solution; these results are omitted for brevity.

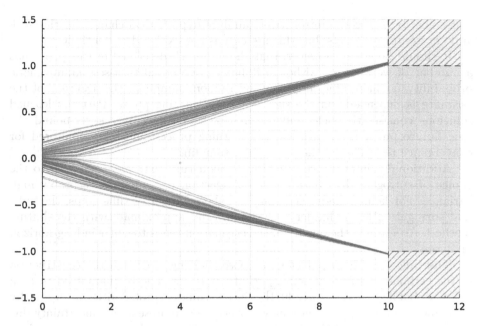

Fig. 3. 100 random evaluations of a failure policy learned with SAC. Actions are instances of the environment that maximize likelihood while eliciting failure: these trajectories are approximately representative of the likeliest failures. Note that the policy encodes the location of both failure modes in its "instructions" for selecting adversarial environments. In the importance sampling scheme, this latent representation is extracted and used to calculate failure statistics.

the next section; the most immediate solution would be simply to increase the sample size.

7   Discussion and Future Work

We have derived and implemented two significant enhancements to adaptive stress testing of complex systems: (1) the ability to learn failure policies that generate the likeliest sequence of environments leading to system failure from arbitrary initializations and (2) the use of these policies in importance sampling to generate statistics that show how likely these failure modes are to occur naturally. These improvements make AST a more powerful tool for failure generation and analysis.

At a high level, these benefits are due to the fact that the model is dissociated during the learning phase (i.e., used descriptively but not generatively) and reassociated in a separate analysis phase. As a result of this separation, there are very few constraints on the solving method: the analysis can proceed from any solution that produces a valid policy. The flexibility also extends to the analysis, which can in theory admit *any* model. The importance sampling scheme can just as easily be used to retroactively analyze hypothetical changes to the SUT and environment models, but this use case is not explored in depth here.

It should be noted, however, that the approach described in this paper has a number of weaknesses. Although it has the potential to assess failure mode probability far more efficiently than by random sampling, the accuracy of the estimate is dependent on the success of the learning phase. Partially learned solutions consistently yield underestimates, which may only lower-bound the true failure probability. As such, the learning process must be monitored for convergence regardless of the subsequent sampling.

Additionally, importance sampling is, as a rule, extremely sensitive to the choice of surrogate distribution; a poor selection can yield unacceptably high variance. For failure modes spanning a higher number of time steps, the optimal surrogate $q^*(x)$ is harder to systematically approximate with the shifting method. As a result, the quality of the estimation tends to suffer in long-horizon analyses.

The range of the output can also pose challenges: though the construction of the estimate ensures its non-negativity, it is not strictly guaranteed to be a valid probability in the range $[0, 1]$. Since estimated probabilities are typically very low, the variance estimate may not accuracy represent the uncertainty due to the inherent skew of the distribution; this effect can be seen in the previous example.

These various issues can compound to produce a situation where the upper confidence limit is inaccurate because of underestimation caused by poor learning or surrogate selection, while the lower confidence limit is not particularly useful because the spread encompasses too many orders of magnitude between the estimate mean and zero. In such cases, the lower limit can be improved with more sampling while the upper limit can be taken from the alternative bounds described earlier.

Many of these issues may be resolved by forming the estimate directly in log-probability space and sampling the optimal surrogate exactly via Monte Carlo Markov chain (MCMC) methods. This is the subject of ongoing research.

8   Conclusion

Adaptive stress testing can be an useful component in the validation of complex stochastic systems. We have addressed several limitations of the standard adaptive stress testing formulation. By generalizing AST results and learning failure policies, we gain the ability to query the likeliest path to failure from any initial state. This enables procedural generation of failures without additional training, which may be useful for diagnostics or runtime assurance tools.

We also show how the learned failure policy can form the basis of an importance sampling scheme to calculate the probabilities of entire failure modes along with probabilistic bounds. We demonstrate that for rare failures, AST with resampling vastly outperforms the standard Monte Carlo method and offers a degree of validative independence that an expert-guided search might lack.

Acknowledgments. This work is supported by the Systems-Wide Safety (SWS) Project under the NASA Aeronautics Research Mission Directorate (ARMD) Airspace Operations and Safety Program (AOSP).

References

1. Corso, A., Du, P., Driggs-Campbell, K., Kochenderfer, M.J.: Adaptive stress testing with reward augmentation for autonomous vehicle validation. In: 2019 IEEE Intelligent Transportation Systems Conference (ITSC), pp. 163–168. IEEE, Auckland (2019)
2. Gleave, A., Dennis, M., Wild, C., Kant, N., Levine, S., Russell, S.: Adversarial policies: attacking deep reinforcement learning. arXiv preprint arXiv:1905.10615 (2019)
3. Goodfellow, I.J., Shlens, J., Szegedy, C.: Explaining and harnessing adversarial examples. arXiv preprint arXiv:1412.6572 (2014)
4. Haarnoja, T., Zhou, A., Abbeel, P., Levine, S.: Soft actor-critic: Off-policy maximum entropy deep reinforcement learning with a stochastic actor. In: International Conference on Machine Learning, pp. 1861–1870. PMLR, Stockholm (2018)
5. Julian, K.D., Lee, R., Kochenderfer, M.J.: Validation of image-based neural network controllers through adaptive stress testing. In: IEEE International Conference on Intelligent Transportation Systems (ITSC), pp. 1–7. IEEE, Rhodes (2020)
6. Kochenderfer, M.J.: Decision Making Under Uncertainty: Theory and Application. MIT Press, Cambridge (2015)
7. Koren, M., Alsaif, S., Lee, R., Kochenderfer, M.J.: Adaptive stress testing for autonomous vehicles. In: IEEE Intelligent Vehicles Symposium (IV), pp. 1–7. IEEE, Changshu (2018)
8. Koren, M., Kochenderfer, M.J.: Adaptive stress testing without domain heuristics using go-explore. In: 2020 IEEE 23rd International Conference on Intelligent Transportation Systems (ITSC), pp. 1–6. IEEE, Rhodes (2020)
9. Koren, M., Nassar, A., Kochenderfer, M.J.: Finding failures in high-fidelity simulation using adaptive stress testing and the backward algorithm. In: 2021 IEEE/RSJ International Conference on Intelligent Robots and Systems (IROS), pp. 5944–5949. IEEE, Prague (2021)
10. Lee, R., Kochenderfer, M.J., Mengshoel, O.J., Silbermann, J.: Interpretable categorization of heterogeneous time series data. In: International Conference on Data Mining (SDM), pp. 216–224. SIAM, San Diego (2018)
11. Lee, R., Mengshoel, O.J., Agogino, A.K., Giannakopoulou, D., Kochenderfer, M.J.: Adaptive stress testing of trajectory planning systems. In: AIAA SciTech, Intelligent Systems Conference (IS), pp. 1454. AIAA, San Diego (2019)
12. Lee, R., et al.: Adaptive stress testing: finding likely failure events with reinforcement learning. J. Artifi. Intell. Res. **69**, 1165–1201 (2020)
13. Lipkis, R., Lee, R., Silbermann, J., Young, T.: Adaptive stress testing of collision avoidance systems for small uass with deep reinforcement learning. In: AIAA SciTech 2022 Forum, pp. 1854. AIAA, San Diego (2022)
14. Mitzenmacher, M., Upfal, E.: Probability and computing: Randomization and probabilistic techniques in algorithms and data analysis. Cambridge University Press, Cambridge (2017)
15. Moss, R.J., Lee, R., Kochenderfer, M.J.: Adaptive stress testing of trajectory predictions in flight management systems. In: IEEE/AIAA Digital Avionics Systems Conference (DASC), pp. 1–10. IEEE, San Antonio (2020)

16. Ng, A.Y., Harada, D., Russell, S.: Policy invariance under reward transformations: Theory and application to reward shaping. In: ICML, vol. 99, pp. 278–287. ICML, Bled (1999)
17. Owen, A.B.: Monte Carlo theory, methods and examples. Preprint, online (2013)
18. Ramakrishna, S., Luo, B., Kuhn, C.B., Karsai, G., Dubey, A.: Anti-carla: an adversarial testing framework for autonomous vehicles in carla. In: 25th International Conference on Intelligent Transportation Systems (ITSC), pp. 2620–2627. IEEE, Macau (2022)
19. Sharif, A., Marijan, D.: Adversarial deep reinforcement learning for trustworthy autonomous driving policies. arXiv preprint arXiv:2112.11937 (2021)
20. Tuncali, C.E., Fainekos, G., Ito, H., Kapinski, J.: Simulation-based adversarial test generation for autonomous vehicles with machine learning components. In: 2018 IEEE Intelligent Vehicles Symposium (IV), pp. 1555–1562. IEEE, Changshu (2018)
21. Watkins, C.J., Dayan, P.: Q-learning. Mach. Learn. **8**(3), 279–292 (1992)

Anxiety Among Migrants - Questions for Agent Simulation

Vivek Nallur(✉)

School of Computer Science, University College Dublin, Dublin-4, Ireland
vivek.nallur@ucd.ie

Abstract. This paper starts with hypothesis (and presents some evidence) that anxiety in migrants is sufficiently important to be modelled. It presents a small (and very incomplete) review of emotion modelling in literature. It asks the question of how to translate these into agent-based modelling, and whether this can be orthogonal to *specific* modelling of goals and capabilities of agents. This short paper is offered as a motivator for discussion, rather than a discussion of results.

Keywords: Economic Migrants · Capabilities Approach · Simulation

1 Introduction

Migration is a complex phenomenon. A number of factors, personal, economic, (geo-) political, cultural, and environmental, are involved in the decision to move, and integrate with, a new society. The 'new' society need not be a different country, with a different language. Even a different physical region within the same country, with different environment, pace of life, urbanization-level, etc. can contribute to a feeling of otherness. If not alleviated, this feeling of otherness can result in anxiety [11]. At a sufficiently large-scale, anxiety can result in deleterious consequences for an individual's physical and mental health. This paper attempts to look at modelling of emotions, specifically anxiety, as well as modelling its spread among migrants. It asks whether a good computational modelling of emotions in agents exists, and if not, what properties it ought to have.

Our investigation into anxiety modelling occurs in the context of the COTHROM[1] project. The COTHROM project attempts to model the post-migration experience of economic migrants. The phenomenon of settling down in a new place, acquiring social capacities, integrating into social networks, etc. is partly a cognitive experience, that has no direct physical, or environmental measure. Hence, we needed to create a model that reflected both, the legal processes that migrants went through, as well as *anxiety* that migrants feel while reacting to these processes. The feeling of being integrated with society is a nebulous one, and there is little data available on any systematic way to measure it.

---

[1] Supported by the Irish Research Council via Grant COALESCE/2021/4.

F. Amigoni and A. Sinha (Eds.): AAMAS 2023 Workshops, LNAI 14456, pp. 141–150, 2024.
https://doi.org/10.1007/978-3-031-56255-6_8

There have been agent-based simulation of migrants previously [8,26,30]. However all of these have been with regard to distressed peoples or refugees, and hence their motivations are sufficiently different from economic migrants, so as not to be directly comparable.

2   Modelling

The presence of anxiety in migrants at a significantly higher rate, than the background/normal rate of the population, has been well-documented [1,5,15]. One might intuitively expect the stress of cultural adaptation to be present in across-country migrants. However, increased rates of anxiety, and even clinical depression, have also been reported among within-country migrants [11], where one would not expect cultural, or language barriers to be stressors. With factors such as climate change, and lack of population growth (in certain regions) making migration an almost-inevitable phenomenon, it is imperative that we understand how to model anxiety in migrants, how it spreads, how it dissipates, and what can be done to alleviate it.

2.1   Modelling Emotions

There are many theories on the nature of emotion, its causation and generation, from James's "affective warmth" [9], to Schachter and Singer's Cognition-Arousal theory [22] to Lazarus's Appraisal theory [25] and many more. While a review of these theories is out of scope for this paper, the common thread (according to this author) running through these theories, is the simultaneous presence of multiple components, with varying levels of activation and inter-relatedness. The precise level of physical (or neuro-physical) stimulus required for a full recognition of emotion remains contested. However what is not contested is the effect of the prior personality. That is, different people may be primed to experience different emotions, in the presence of the same stimuli, depending on their expectation from the context [2].

2.2   Classical Theory of Emotions

In affective computing, the sub-field of sentiment categorization and analysis, is concerned with the attempts to infer the emotions being expressed or experienced, given the state of some perceptual variables. One of the earliest attempts to define a model began with Shaver*et al.* [24], who first categorized words into emotion words and non-emotion words. Then, through an iterative process of clustering and hierarchy generation, reached a 'basic' level which contained six emotions: *anger, fear, love, joy, sadness* and *surprise*. Subsequent research has contested this notion of basic emotions and proposed multiple other models, such as the Valence-Arousal model [14], revised OCC model [19], Hourglass model [4], the revised hourglass model [28], the Ekman model [6], and Plutchik's Wheel of Emotions model [21], among others. The Wheel of Emotions uses an evolutionary model as the foundational explanatory framework, and posits that emotions coexist with functional abilities, as an essential adaptive mechanism.

The Wheel of Emotions has been fairly popular in computational modelling. Perhaps this is so (in the author's opinion), since it offers a continuous three-dimensional model which allows functions to work upon a particular state variable, and either categorize or predict the result. Other models that offer discrete categories may be simpler, however there is no guidance (from a computational point of view) as to how a variable moves from one state to the other. Of particular interest is Gratch and Marsella's work on a domain-independent computational interpretation of appraisal theory [7]. This theory (barbarically summarized) views emotions as emerging out of two basic processes, *appraisal* (evaluating the significance of events), and *coping* (strategies for responding to the appraisal). The claim of the theory is that causal interpretation of the appraisal variables combined with adaptable plan representations and world models are able to provide "a first approximation of the type of reasoning that underlies appraisal and coping". Figure 1 shows the various components involved in implementing such a theory.

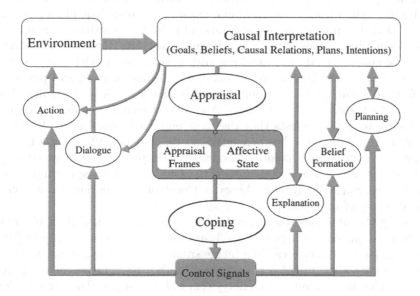

Fig. 1. A computational instantiation of appraisal theory. From Gratch and Marsella [7]

2.3 Affective Neuroscience

In the classical theory, emotions are basic to human beings and cannot be broken down further. However, some scholars in the field of neuroscience have a different view. The Theory of Constructed Emotion [2] as proposed by Lisa Feldman Barrett, for instance, posits that what we recognize as emotion is actually a complex interplay of past experience, cultural cues, and interoception. This, according to the theory, has better explanatory power, and is supported by empirical evidence, as opposed to the classical psychophysiological theory. It would be overstating the case, to imply that this is well-accepted even among neuroscientists. While scholars such as Joseph LeDoux [12] seem to agree with Barrett, however others such as Panksepp disagree [20]. According to Panksepp [20], Barrett disregards the available evidence from cross-species neuroscience, which suggests that interoceptive systems of visceral brains are involved in generation of affective properties. LeDoux, on the other hand, agrees with Barrett that while subcortical circuits in the brain are responsible for detecting and responding to sensory perceptions, their activity do not constitute emotions by themselves [13].

2.4 Computer Science

Agent-based modelling has been used to study patterns in a variety of complex adaptive systems [16,27] as well as to test strategies to induce emergent behaviour [17]. This gives us confidence that using that agent-based modelling could be a good computational approach for understanding the decision-making of migrants. The variation in emotional experience, conditioned by prior expectations as well as precise local conditions, all point to agent-based simulation as a feasible mechanism for modelling migrant behaviour. However, modelling the impact of emotion to decision-making such agent-based models have been less than conclusive. The lack of consilience between neuroscience, and psychology, means that researchers in computer science have to pick a particular theory, based on their own modelling needs. A review of emotion modelling in agent-based simulation [3] suggests that the most popular theoretical model adopted by various researchers has been the classical OCC model [19]. Even a literature survey on computational emotion modelling (without regard to social simulation) [10] appears to confirm this, with the OCC model being the most widely implemented. Surprisingly, the Theory of Constructed Emotion is not even referenced in any of these surveys.

Among the literature that specifically addresses the modelling of anxiety, Vanhée et al. propose using *Anxiety-Aware Markov Decision Processes* [29] for representing uncertainty regarding future situations, and their associated probabilistic rewards, also grounded in psychological theories. The psychological (classical) theories on emotion and anxiety seem to be most popular, perhaps due to the ease (we use this word in a relative sense) of integrating it with regular computational functions. In this paper, we do not take a position on which theory would be best to model anxiety, specifically.

3 Our Currrent Agent Model

In our approach, we are inspired by Sen [23] and Nussbaum's capability model [18] to inform our notion of what a migrant should be able to do. According to the capabilities approach, a human being may be seen not just as *being* but also as a set of *"doings"* and *"functionings"*. That is, the quality of life being led by a human being, can be seen as evaluating the *functionings* and the *capability to function*. By definition, this indicates that there are several dimensions involved in experiencing and evaluating the capability to function. Hence, we decided to define a migrant's state of life in the new host country, as a function of the number of capabilities that they are able to acquire. Our *assumption* is that the quicker a migrant is able to acquire a verisimilitude of capabilities as that of a citizen, the more satisfied they are, at having migrated. There are a range of capabilities that an average adult citizen is able to wield. We attempt to capture some of these:

1. **Labour Mobility**
2. **Access to Health Services such as GPs**
3. **Access to Maternity Care**
4. **Access to Child Care**
5. **Access to Rented Housing**

We are acutely aware that these do not form the entirety of the lived experience of a migrant. However, we would like to ensure that our simulation's input data have internal validity in the form of macro-economic data, that is source-able from reputable, third-party sources. A different, yet important, form of validation for our model's results are its consonance with trends from personal interviews also being conducted as a part of our work. However, since these have not yet been completed, we do not mention them further in this paper.

3.1   The Multi Queue Model Approach

The multi-queue approach imagines a migrant (and their family) as being present in multiple queues, some due to legal processes and others due to their own physical, social and emotional needs. These queues form an approximation of state capacity in providing access to resources. These resources may be of multiple types: *legal* (e.g., access to a PPSN[2]), *physical* (e.g., access to housing), *socioeconomic* (e.g., being accepted in a GP surgery). The `arrival rate` for the queue associated with each resource is a function of the number of agents trying to access it, while the `service time` is a function of the type of resource. That is, certain kinds of legal resources have a well-defined service time (e.g., the mean service time, as measured by the Central Statistics Office, to be allocated a PPSN is 42 days), whereas others (e.g., being accepted in a GP surgery) have a service time that is sourced from media reports, and migrant interviews. The service time is used in the simulation to move a migrant(agent) through time, and allows us to visualize how many agents possess what kinds of capabilities, as time passes.

The capabilities that the agents seek to acquire, are not affected *solely* by the economy, and the social framework they find themselves in. They are also impacted by historical factors, which may sometimes be somewhat measurable

---

[2] A PPSN is a number that is allocated to every adult individual and is used for multiple purposes, from taxation to vaccination.

and also time-varying. The phrase *somewhat measurable* is used after some consideration. `Wealth`, for instance, can be measured as a function of the estimated income received by the migrant in their new job. However, economic migrants may also have access to savings or family-based wealth, which affects their actual disposable income. A further variable that affects `Wealth` is `Spouse Employment` which is further impacted by laws that regulate whether a spouse has an automatic right to work, or must follow a different process to be allowed gainful employment. A further wrinkle is an employment law condition that certain migrants (of certain income-groups, and/or certain professions) must remain with the same employer for a specified time-period, for their work-visa to remain valid. If the employer terminates employment within this period, or the employee stops working with the employer, then the visa condition is violated, and the migrant's permission to remain may be cancelled. These constraints on capabilities are modelled as lack of capabilities (e.g., labour mobility) when benchmarked against the background level of capabilities that typical citizens enjoy. It is pertinent to note that these capabilities change with time, as different conditions have a *knock on* effect on the initial starting conditions. An example would be about the capability to access health services. Although the law does not forbid access to healthcare, the lack of labour mobility (affecting `Wealth`) may affect a migrant's ability to reside in a location, which may further affect their access to GP services. The stress brought upon by lack of access to healthcare is denoted (partially) by the `Health` variable. These variables interact with one another in various ways, denoting either increase in capabilities or the lack of them. The simulation names the composite value of capabilities across the various factors as `Satisfaction`. The delays suffered due to interaction between multiple dimensions also affect the anxiety level of the migrants.

4 Technical Details

We use `Mesa`, a Python-based library to build our agent models.[3] Our key agent is an immigrant with a work permit. To create representative agents, we first construct a sample of immigrants using various data on immigrants in Ireland. Table 1 reports the details of the variables and sources from which we retrieved the data. The Department of Justice (DoJ) has provided us with demographic data on non-EU immigrants with work permits in Ireland. According to the data from the DoJ, the number of non-EU immigrants who have been granted work permits between 2019 and 2021 is about 19,000.[4] From the constructed sample, we randomly draw 5,000 agents for our model. The code for the model and the analysis can be freely downloaded from our repository.[5]

In the model, each agent seeks to achieve a certain level of satisfaction along seven dimensions: economic wealth, housing, health care, employment (visa renewal), spouse employment, maternity care, and child care. The number of eligible layers depends on the immigrants' family status. The first four dimensions

---

[3] https://mesa.readthedocs.io.

[4] But only about 9,500 immigrants have information on their positions.

[5] https://github.com/viveknallur/cothrom-mabs-2023.git.

are applicable to all immigrants, whereas the last three dimensions are applicable only to those having partners and/or their children.

To measure each immigrant's overall satisfaction, we construct the satisfaction composite index. This index calculates the proportion of the layers in which they are satisfied to the number of all layers they want to achieve satisfaction.[6] The value of the composite index is defined as **one**, if the immigrant is fully satisfied in all possible dimensions.

The unit of time in our model is a day. Everyday, each immigrant performs some goal selection, *i.e.,* they pick a task they seek to complete, including applying for PPSN and residential permit, getting an apartment rent or buying a house, visiting a GP if needed, applying for recognition of his/her partner's work experience or education, and applying for a new position, etc. The amount of energy they are able to expend on this task is a function of existing measurable variables, such as `Wealth` and `Health`, and also the anxiety they experience with respect to that goal. Immigrants in the model can be involved in at maximum two different tasks within a day. If they successfully accomplish the task, they are assumed to be satisfied in that dimension.

Table 1. Data Sources and Variables

Variables	Measurement (Source)
Demographic data (origin country, position, age, gender, and marriage)	Department of Justice
Initial wealth	GDP per capita (World Bank)
English fluency	UN Statistics Division & CIA Factbook
Salary	Morgan McKinley[a]
Education level	CareersPortal.ie[b]
Housing market	Rent report & Price of Houses (Central Statistics Office[c])

[a]https://www.morganmckinley.com
[b]https://careersportal.ie
[c]https://data.cso.ie

4.1 Adding Anxiety to the Agent Model

While the computational aspects of this simulation work well, there are several aspects of the model that are not ideal. Measuring how welcome a migrant feels, and how quickly and thoroughly they are able to acquire capabilities that allow them to live a full life, is a difficult task. The current state-variable approach is **not a true representation of the cognitive state of a migrant**. Anxiety is both, an individual state, as well as diffuses through a social network. This can

---

[6] For `wealth` dimension, we define as satisfied if the migrant's cumulative wealth values are greater than his/her half of annual salary.

have real consequences for goal selection, as well as for the amount of energy employed in goal achievement. Anxiety works in both positive, as well as negative ways. Mild amounts of anxiety on a particular dimension, result in that particular goal-task being selected multiple times. Non-satisfaction of the goal, either due to variables associated with the individual agent, or due to environmental factors over which the agent has no control, leads to increased anxiety. The stressor variables that we would like to include are:

- Lack of recognition of educational/practice credentials
- Family relocation
- Religion/ethnicity based discrimination
- Incivility at workplace
- Socio-cultural adaptation stress

4.2   Applying the Computational Models

As per the computational models discussed in Sect. 2.2, the fundamental requirement for modelling emotions is to maintain a state-based mapping of goals, capabilities and then evaluate the likelihood of success/failure of plans to achieve goals. However, there seems to be no obvious way to model the anxiety the stressors variables induce (e.g., incivility at workplace or religion based discrimination), apart from introducing *extremely* higher-level proxy goals (e.g., pleasant social interaction). The proxy goals would then require satisfaction processes of their own, which threaten an infinite regress of proxies. Social influences of community support structures, family and friends are also ignored in these models. Given this context, we ask the following questions:

1. What is a viable computational model for modelling an emotion that is not a direct consequence of invisible/proxy variables? Can the same emotion be used for positive and negative effects?
2. How do we validate a computational implementation of emotion, from macro third-party data?
3. How do social influence dynamics affect computational models of emotions?
4. Should resource and task modelling be independent of (or orthogonal to) emotion modelling? If not, how do we retain comprehensibility, in terms of architectural/design patterns?

5   Conclusion

We are acutely aware that the simulation is a pale reflection of the cognitive state of a migrant, and a coarse proxy of the interaction of various migration laws present in Ireland, and their (possibly inadvertent) interaction with domestic capacities, and constraints. The simulation is nevertheless, we believe, a better mechanism for understanding the impact of policy change, with differentiated results on different migrant sub-communities. A majority of the team-members of COTHROM are migrants themselves, and hence are able to reflect on their own experiences to ensure that the simulation, while striving to be objectively faithful to recorded macro-level data, is also able to surface micro-level problems, that are currently invisible due to lack of representation. The simulation may be used as a tool to consider the cost-benefit impact of policy changes, and (hopefully) to prevent any well-intentioned policy from having a deleterious impact.

References

1. Aalto, A.M., et al.: Employment, psychosocial work environment and well-being among migrant and native physicians in Finnish health care **24**(3), 445–451. https://doi.org/10.1093/eurpub/cku021
2. Barrett, L.F.: The theory of constructed emotion: an active inference account of interoception and categorization **12**(1), 1–23. https://doi.org/10.1093/scan/nsw154
3. Bourgais, M., Taillandier, P., Vercouter, L., Adam, C.: Emotion modeling in social simulation: a survey **21**(2), 5. https://doi.org/10.18564/jasss.3681. https://jasss.soc.surrey.ac.uk/21/2/5.html
4. Cambria, E., Livingstone, A., Hussain, A.: The hourglass of emotions. In: Esposito, A., Esposito, A.M., Vinciarelli, A., Hoffmann, R., Müller, V.C. (eds.) Cognitive Behavioural Systems. LNCS, vol. 7403, pp. 144–157. Springer, Heidelberg (2012). https://doi.org/10.1007/978-3-642-34584-5_11
5. Cayuela, A., Malmusi, D., López-Jacob, M.J., Gotsens, M., Ronda, E.: The impact of education and socioeconomic and occupational conditions on self-perceived and mental health inequalities among immigrants and native workers in Spain **17**(6), 1906–1910. https://doi.org/10.1007/s10903-015-0219-8
6. Ekman, P.: An argument for basic emotions. Cogn. Emot. **6**(3), 169–200 (1992)
7. Gratch, J., Marsella, S.: A domain-independent framework for modeling emotion **5**(4), 269–306. https://doi.org/10.1016/j.cogsys.2004.02.002. https://www.sciencedirect.com/science/article/pii/S1389041704000142
8. Hébert, G.A., Perez, L., Harati, S.: An agent-based model to identify migration pathways of refugees: the case of Syria. In: Perez, L., Kim, E.-K., Sengupta, R. (eds.) Agent-Based Models and Complexity Science in the Age of Geospatial Big Data. AGIS, pp. 45–58. Springer, Cham (2018). https://doi.org/10.1007/978-3-319-65993-0_4
9. James, W.: Principles of Psychology, 2nd edn. Holt, New York (1939)
10. Kowalczuk, Z., Czubenko, M.: Computational approaches to modeling artificial emotion - an overview of the proposed solutions **3**. https://www.frontiersin.org/articles/10.3389/frobt.2016.00021
11. Lam, K.K.F., Johnston, J.M.: Depression and health-seeking behaviour among migrant workers in Shenzhen **61**(4), 350–357. https://doi.org/10.1177/0020764014544767
12. LeDoux, J.E.: The emotional brain: the mysterious underspinnings of emotional life. Simon & Schuster Paperbacks, 16th print edn. OCLC: 605049863
13. LeDoux, J.E., Brown, R.: A higher-order theory of emotional consciousness **114**(10), E2016–E2025. https://doi.org/10.1073/pnas.1619316114, https://www.pnas.org/doi/10.1073/pnas.1619316114. Publisher: Proceedings of the National Academy of Sciences
14. Mehrabian, A.: Pleasure-arousal-dominance: a general framework for describing and measuring individual differences in temperament. Curr. Psychol. J. Diverse Perspect. Diverse Psychol. Issues **14**, 261–292 (1996)

15. Meyer, S.R., Decker, M.R., Tol, W.A., Abshir, N., Mar, A.A., Robinson, W.C.: Workplace and security stressors and mental health among migrant workers on the Thailand-Myanmar border **51**(5), 713–723. https://doi.org/10.1007/s00127-015-1162-7

16. Nallur, V., Cardozo, N., Clarke, S.: Clonal plasticity: a method for decentralized adaptation in multi-agent systems. In: Proceedings of the 11th International Symposium on Software Engineering for Adaptive and Self-Managing Systems, pp. 122–128. ACM. https://doi.org/10.1145/2897053.2897067. https://dl.acm.org/doi/10.1145/2897053.2897067

17. Nallur, V., O'Toole, E., Cardozo, N., Clarke, S.: Algorithm diversity: a mechanism for distributive justice in a socio-technical MAS. In: Proceedings of the 2016 International Conference on Autonomous Agents & Multiagent Systems, pp. 420–428 (2016)

18. Nussbaum, M.C.: Creating Capabilities: The Human Development Approach, Reprint edition. Harvard University Press, Cambridge (2014)

19. Ortony, A., Clore, G.L., Collins, A.: The Cognitive Structure of Emotions. Cambridge University Press, Cambridge (1988)

20. Panksepp, J.: Neurologizing the psychology of affects: how appraisal-based constructivism and basic emotion theory can coexist **2**(3), 281–296. https://doi.org/10.1111/j.1745-6916.2007.00045.x. Publisher: SAGE Publications Inc

21. Plutchik, R.: The nature of emotions: human emotions have deep evolutionary roots, a fact that may explain their complexity and provide tools for clinical practice **89**(4), 344–350. https://www.jstor.org/stable/27857503. Publisher: Sigma Xi, The Scientific Research Society

22. Schachter, S., Singer, J.: Cognitive, social and physiological determinants of emotional states. Psychol. Rev. **69**, 379–399 (1962)

23. Sen, A.: Development as capability expansion. In: Readings in Human Development, pp. 2–16. 1.1. Oxford University Press

24. Shaver, P., Schwartz, J., Kirson, D., O'connor, C.: Emotion knowledge: further exploration of a prototype approach. J. Pers. Soc. Psychol. **52**(6), 1061 (1987)

25. Smith, C.A., Lazarus, R.S.: Emotion and adaptation. In: Pervin, L.A. (ed.) Handbook of Personality: Theory and Research, pp. 609–637. The Guilford Press, New York (1990)

26. Sokolowski, J.A., Banks, C.M.: A methodology for environment and agent development to model population displacement. In: Proceedings of the 2014 Symposium on Agent Directed Simulation, ADS 2014, pp. 1–11. Society for Computer Simulation International (2014)

27. Song, H., Elgammal, A., Nallur, V., Chauvel, F., Fleurey, F., Clarke, S.: On architectural diversity of dynamic adaptive systems. In: 2015 IEEE/ACM 37th IEEE International Conference on Software Engineering, vol. 2, pp. 595–598. https://doi.org/10.1109/ICSE.2015.201. tex.organization: IEEE

28. Susanto, Y., Livingstone, A.G., Ng, B.C., Cambria, E.: The hourglass model revisited **35**(5), 96–102. https://doi.org/10.1109/MIS.2020.2992799. https://ieeexplore.ieee.org/document/9237283/

29. Vanhée, L., Jeanpierre, L., Mouaddib, A.I.: Anxiety-sensitive planning: from formal foundations to algorithms and applications **32**, 730–740. https://doi.org/10.1609/icaps.v32i1.19863. https://ojs.aaai.org/index.php/ICAPS/article/view/19863

30. Warnke, T., Reinhardt, O., Klabunde, A., Willekens, F., Uhrmacher, A.M.: Modelling and simulating decision processes of linked lives: an approach based on concurrent processes and stochastic race **71**, 69–83. https://doi.org/10.1080/00324728.2017.1380960

Should My Agent Lie for Me? Public Moral Perspectives on Deceptive AI

Stefan Sarkadi[1][(✉)], Peidong Mei[2], and Edmond Awad[2]

[1] King's College London, London, UK
stefan.sarkadi@kcl.ac.uk
[2] University of Exeter, Exeter, UK
{p.mei,e.awad}@exeter.ac.uk

Abstract. Artificial Intelligence (AI) advancements might deliver autonomous agents capable of human-like deception. Such capabilities have mostly been negatively perceived in HCI design, as they can have serious ethical implications. However, AI deception might be beneficial in some situations. Previous research has shown that machines designed with some level of dishonesty can elicit increased cooperation with humans. This raises several questions: Are there future-of-work situations where deception by machines can be an acceptable behaviour? Is this different from human deceptive behaviour? How does AI deception influence human trust and the adoption of deceptive machines? In this paper, we describe the results of a user study published in the proceedings of AAMAS 2023. The study answered these questions by considering different contexts and job roles. Here, we contextualise the results of the study by proposing ways forward to achieve a framework for developing Deceptive AI responsibly. We provide insights and lessons that will be crucial in understanding what factors shape the social attitudes and adoption of AI systems that may be required to exhibit dishonest behaviour as part of their jobs.

Keywords: Deceptive AI · AI Ethics · User Study

1 Introduction

It is A's object in the game to try and cause C to make the wrong identification. [...] 'What will happen when a machine takes the part of A in this game ?' [79]

Deception has many definitions and comes in numerous forms [47], but it is universally considered to be the process through which one entity causes another entity to have a false belief [48,62]. In the area of computing, Alan Turing was the first to give deception a most special role, namely that of indicating the answer to the most fundamental questions about computers, which is *'Can machines*

This project was supported by the Royal Academy of Engineering and the Office of the Chief Science Adviser for National Security under the UK Intelligence Community Postdoctoral Research Fellowship programme.

F. Amigoni and A. Sinha (Eds.): AAMAS 2023 Workshops, LNAI 14456, pp. 151–179, 2024.
https://doi.org/10.1007/978-3-031-56255-6_9

think?' 72 years after Turing's reflections on the relation between machine intelligence and deception, the topic of deception is becoming more prevalent than ever in the debate about Artificial Intelligence (AI) and society. But, while most focus has been on the immediate risks posed by online AI-enabled deception, such as the intentional propagation of fake news with the aid of trollbots or DeepFakes, research on the future risks coming from the advancements of fully autonomous deceptive agents, as imagined by Turing in his imitation game, has been scarce [62]. In addition to this, more than 20 years have passed since Castelfranchi's futuristic argument that artificial liars were a natural development in virtual societies and explained *'why computers will (necessarily) deceive us and each other'* [10]. However, Castlefranchi's vision was not just about the necessity of malicious deception, but also about the self-interest of deceptive autonomous agents which would aim to achieve pro-social goals. That is AI agents which deceive for both their own benefit and that of humans. According to [10,11], deception comes in several forms of interaction between humans and machines, namely 1) the agent deceives for its principal: the mandatary deceives through its agent; 2) the agent deceives autonomously; and 3) the agent deceives its own principal/user.

These human-AI interactions are also emphasised in the context of future-of-work, where AI agents will replace some roles humans currently perform, such as self-driving cars. Research at the intersection of AI ethics and the future-of-work offer us insights into the emergence of possible social norms, e.g. the Moral Machine Experiment offers us insights as to what humans would think is the right thing to do for a self-driving car in trolley-problem style scenarios [2]. This allows AI researchers to think about how to design the machines that would fill the roles of humans to behave according to a certain society's norms, e.g. do what the humans of that society would.

What has not yet been explored in the literature, is whether in some of these future-of-work job roles machines are expected to deceive. Humans deceive according to the norms that apply to them in specific contexts. Could machines of the future do the same? To build AI agents that follow social norms when attempting deception, we must first find out if it is possible to build an ethical normative framework for this. Hence, in order to find out if such a framework can be built, our aim in this paper is to understand empirically the relation between humans and machines in 5 different contexts where autonomous AI agents deceive for the benefit of humans. In this paper, we provide answers to the following research questions w.r.t. the 5 selected contexts:

RQ1 When would AI deception be perceived as more permissible by humans?

RQ2 Would humans trust AI agents capable of deception?

RQ3 Would humans want to adopt or buy AI agents capable to deceive ?

RQ4 Who would humans hold responsible, the deceiver, the beneficiary, or, in the case of AI deception, the designers of the AI-powered machine?

To answer these questions, we have designed a survey to elicit user feedback on 5 scenarios/stories involving an agent (human or machine powered with AI technology) fulfilling different roles in which it deceives for the benefit of another entity (another human individual, or an organisation). According to Castelfranchi's interaction types, this would be type 2, where the AI agent deceives autonomously and deliberately, but for the benefit of its principal (as in type 1) [10]. To clarify, we are talking here about studying deliberate and intentional deception for both humans and machines during these interactions. The lack of intentionality, however, is also addressed in our study, by introducing the roles of beneficiary of deception that is different from the deceiver, and the role of an AI designer/maker, and, later in the paper, linking these roles to the literature the ethical norms to govern deceptive behaviour of machines, such that they do not perform unintentional deception due to their irresponsible design specifications [24].

We define the primary hypotheses regarding moral permissibility, responsibility, and willingness to buy w.r.t. dishonest behaviour by AI-powered machines:

Hypotheses

H1.M: Agent types, deception roles, and demographic differences affect how people assign moral permissibility to deceptive agent behaviour.

H2.T: Agent types, deception roles, and demographic differences affect how much people trust deceptive agents.

H3.W: Agent types, deception roles, and demographic differences affect how willing people are to buy the services of deceptive agents.

H4.R: Agent types, deception roles, and demographic differences affect how people assign responsibility to the entities involved in the deception.

---

2 Background and Related Work

We clearly need a socio-cognitive computational theory of trust and deception as argued in [11,12,28,29]. While the influence of trust in human-AI interactions has been extensively researched, deception has been mostly set aside. Despite this shortcoming, the literature points out that research at the intersection of AI and deception is crucial for several very good reasons [69], such as: (i) to prevent machines from employing malicious deception and lead our societies to a Tragedy of The Digital Commons (TDC) [34,68], (ii) to build machines with human-like intelligence and social abilities [62], and (iii) to align deceptive machines to human values (social norms) such that we reap their long-term benefits [41]. The latter reason (iii) motivates this paper.

What are then, the ethical issues that we face regarding reaping the benefits of deceptive AI? A first issue regarding AI ethics is that deceptive AI has a strong impact on persuasive technologies developed for marketing, where deception is a main strategy to persuade individuals for the benefit of others [32]. Even more so, deceptive design can be used for coercing individuals into using a company's software [43]. According to [51] today's deceptive AI has more to do with human perception rather than the oftenly missing 'intentionality' of AI agents [51]. Similarly, [55] introduces the concept of 'banal' deception to explain how humans are highly susceptible to persuasive technologies, emphasising that it is enough for humans to be put in the right context with technologies for deception to happen, even if the technology itself has no intention of deceiving [55]. But future cognitive AI agents would presumably become more advanced than the tools used in persuasive technologies - they might have a higher degree of autonomy. This brings us to the idea of ethical and responsible design not just of HCI (Human-Computer Interaction) systems that are designed by humans (intentionally or not) to deceive for very specific contexts [1], but of AI agents that can deliberately deceive autonomously in different contexts [24].

Can AI deliberately deceive? Several notable works in AI and multi-agent systems show that practical reasoning AI agents can do it. Almost 20 years ago, [23] demonstrated a simulation tool that can apply deliberative deception in Turing's imitation game [23]. More recently, [67] demonstrated how practical reasoning agents can be designed to perform human-like deliberative deception by forming and using Theory-of-Mind under uncertainty during multi-agent communication [57,64,67]. Moreover, [15] demonstrates in a user study that by using the right type of arguments, machines can successfully deceive human adults [15]. So, yes, with the right cognitive architectures and models, AI can deliberately deceive, and we have evidence that, in principle, it could do so successfully. Hence, knowing this, what should we do with these kinds of deceptive AI agents?

Some believe the right way to deal with deceptive AI agents is to design strategies for sand-boxing them [36,86]. Alternatively, others think that deceptive AI can be beneficial and argue that humans might benefit from an ethically aligned deceptive machine [41]. Deliberate AI deception, as opposed to deceptively designed AI, could in fact promote pro-social norms. What [41] argue is that for machines to be able to deceive in an ethical manner, they must be able to distinguish pro-social goals from malicious goals [41]. This is a consequentialist argument, namely that instead of following a deontological set of pre-defined rules which may prohibit deception under any circumstances [44], the machine reasons about the consequences of their deceptions w.r.t. the social norms that apply in different contexts. Apart from being blind to context, a deontological approach to deception could lead to ambiguous ethical deceptions, such as the passive a priori deception, where the deceiver does not even have to actively try to deceive, but merely relies on the erroneous reasoning of its unfortunate target - which would be permissible even according to Kant, who considered lying to be impermissible under any circumstance [76,77]. In other words, the ethical deceptive machine must align itself to the social norms that humans follow and identify the particular contexts in which humans apply them.

This backing argument for developing ethical deceptive AI in [41] is also based on the idea of prosocial lies, which are lies that are used to promote benevolence-based trust between agents of a society and protect the common good, as evidenced in [46]. However, the same study indicates that integrity-based trust is harmed by prosocial lies [46]. Again, these sorts of effects reinforce the argument that trust must be understood in relation to deception.

One way to increase trust in AI is to design virtual agents that provide explanations, which, according to [84], decreases the perception of AI being deceptive.

Inversely, the ability of AI to deceive also plays a crucial role in explainability, where it can be used for educational purposes [74], or for increasing the team-work performance in search-and-rescue scenarios [13]. This is most evident in the study by [42], where deceptive AI-bots are shown to have a significant advantage at inducing human-AI cooperation, whereas AI-bots that disclose their artificial nature perform worse than humans at promoting cooperation [42]. These bene-fits of deceptive AI are increasingly reflected in the HCI community's discussions regarding design principles for human-AI cooperation [83].

The same concerns that now emerge in the HCI community also emerge in robotics [83]. [72] point out that in social robotics, the usual factors of trust, intentionality (or lack of it), and harmful effects emerge when the term deceptive AI is summoned, but emphasises that the ethics of deception should be regarded w.r.t. to the harm inflicted on society, and not w.r.t. the benefit of the deceiver (or the beneficiary of the deception) [72]. For instance, [22] argues that not all forms of AI deception are harmful, but that special attention must be given to AI deception that is perceived as betrayal from an ethical perspective [22]. The message we should take from [72]'s argument is that it is necessary to ensure that deception in social robotics does not lead to AI replacing meaningful human-human interactions, or to misplaced trust in AI-powered machines. To the further benefit of human-AI cooperation, [6] argue that robots could actually use deception to nudge humans into being better social actors [6].

Indeed, robotics has a relatively strong track of studying the effects of decep-tive behaviour in both human-AI and AI-AI interactions. The works of [81] cre-ated a taxonomy of when robots should deceive [81] and then provided robots with an algorithm that enabled them to tell whether or not deception is war-ranted in a social situation [82]. Later on, [25] designed a robot algorithm for deceptive motions based on human motions and studied its effects when humans interact with the robots [25]. A Wizard-of-Oz experiment by [85] showed that children are highly susceptible to robot deception, and have a tendency to assign human-like properties to robots [85]. On the other hand, a longitudinal study, by [80], showed that older adults' attachment to robots is not affected if robots deceptively express 'emotions', but the authors emphasised that the results are not necessarily generalisable due to the small sample and that more research is needed on the effects of emotional deception by robots on human attach-ment [80]. The good thing about these studies is that they are accurate, well designed, and controlled in the lab. The downside is that they are strictly focused on behaviour and very context-specific, which made it difficult to consider the deliberative cognitive reasoning of AI agents. Yet, they are a good example for relevant research at the intersection of AI ethics and deception.

[61] argues, that robots, such as the ones mentioned in the studies above, should be considered more as vessels of deception, rather than deceivers, but more importantly, AI deception should be studied w.r.t. to its effects on a larger scale rather than at an individual level [61]. [61] further argues that both trust between agents and evolutionary pressures in hybrid societies could very well be influenced by AI deception, and that these effects should be studied both from a philosophical and empirical perspective in order to determine whether AI

deception is malicious or prosocial. Indeed, [34] have previously pointed out that deceptive AI agents which do not align themselves to ethical or prosocial values might lead to negative societal outcomes such as the Tragedy of the Digital Commons [34]. Going beyond the philosophical argument, [68] actually run a large scale agent-based simulation to show how malicious deception emerges and destabilises cooperation in hybrid societies where humans and AI agents interact [68]. On a positive note, [68] show in the same study how the presence of a decentralised regulation mechanism helps hybrid societies organise themselves and re-establish cooperation [68]. Later, [63] show how the same evolutionary process might lead to an arms race between deception and its regulation.

Also positively, [16] speculates that in the future the affective robots that humans might consider deceptive now due to their artificial nature, might actually help humans adopt new value systems that will make them feel more secure in social relationships - the only caveat is that deception must be appropriate to the context in which it is being performed by the machine [16]. Another interesting argument for integrating deceptive AI as part of our hybrid societies comes from the benefits of entertainment. [17] describes and evaluates deception from the perspective of magic and storytelling [17]. Again, context and consequentialism prove to be crucial concepts for classifying wether deception is malicious or prosocial. [17] emphasises that deception is a co-performance whose morality is guaranteed only when the values and expectations of the agents involved in the co-performance are aligned. The AI agents, the human users, and the eventual designers of these machines can all be co-performers, according to [17], but that the responsibility for deception falls onto the entities who have the capabilities to shape the social structures that define who has the power to deceive or let others perform the deception [17]. For instance, [53] show that humans can be nudged by the ones who control the human-agent interaction to endorse deceptive AI behaviour for their benefit if they are forced to experience beforehand a negative or tough negotiation with an agent [53].

As the literature suggests, to reap the benefits of deceptive AI, our understanding of agent-agent interactions must be relative to the ethical values and social norms that humans apply in various contexts. In this paper, we study how humans perceive such interactions w.r.t. moral permissibility, trust, responsibility, and willingness to buy deceptive services.

3   Methods

Participants. 810 participants were recruited via Amazon Mechanical Turk (MTurk), of which 424 successfully passed the attention checks and completed the test online via Qualtrics. The final sample of 424 participants are residing in the US, aged between 21 - 66 (M = 33, SD = 9), Nfemale = 183 (43.16%), Nmale = 241(56.84%). Their self-rated Socio-Economic Status (SES) on a 11-point scale (0-10) is slightly higher than the midpoint of scale (M = 6.60, SD = 2.30). Measured on similar scales for religiosity (anchored at "Not religious" and "Very religious") and political view ("Progressive" and "Conservative"), participants lean towards religious (M = 6.45, SD = 2.47) and conservative (M = 6.52, SD = 2.47). The majority has indicated having undergraduate education (57.10% had a bachelor's degree or attended universities or colleges). The rest have completed a postgraduate degree (master, PhD, or professional degrees; 21.50%), high school diplomas (18.40%), and 3.10% are on vocational training or did not attend high schools. Income wise, 56.37% of the sample earn a medium level income (ranged from $40,000 to $79,999), 30.42% are with low income of less than $40,000 and 13.21% earn more than $80,000. A detailed summary of the demographic information of the sample is shown in Table 1.

Table 1. Summary of Demographic Sample.

	Mean	SD	Min-Max	
Age (ys)	33.00	9.00	21-66	
Social Economic Status (Worst [0] off to Best off [10])	6.60	2.30	0-10	
Religiosity (Not religious [0] to Very Religious [10])	6.45	2.47	0-10	
Political View (Progressive [0] to Conservative [10])	6.52	2.47	0-10	
Gender (%)	*Male*		*Female*	
	57.84		43.16	
Income (%)	*Low* (< $40k)	*Medium* ($40k< $80k)	*High* (> $80k)	
	30.42	56.37	13.21	
Education (%)	*Vocational or Primary*	*High School*	*Undergraduate*	*Postgraduate*
	3.10	18.40	57.10	21.50

Procedures and Study Design. Before starting the survey, participants were required to read a detailed introduction of the study and provided consent if they wanted to take part. This study was approved by the University of Exeter Research Ethics and Governance (REG) Committee. Once consent was given, participants were presented with the survey, which they completed online. The survey took about 10 min on average to finish. Upon successfully completing the survey (without failing any attention checks), each participant was compensated with $1.2 for their time.

The survey consists of 3 parts: 1) the "Deception Judgement" block in which participants make judgements about deceptive behaviours in different contexts; 2) the "Human Intelligence Task (HIT) Questions" block which provides the opportunity for participants to explain their judgements and give feedback; 3) the "Demographics" block which asks for participants' age, gender, education, socio-economic status (SES), religiosity and political view. Please see the Supplementary Materials for a list of all survey questions.

A 2x3 between-subject design was used in the study, specifically for the "Deception Judgement" block. The first factor is the "Dishonest Agent" (human vs. AI). The second factor is "Deception Beneficiary and Target". Depending on whether the subject is the Beneficiary or is the Target, there are 3 levels: 1) deceiving others for the subject's benefit ['Agent4You']; 2) deceiving the subject for others' benefit ['Agent4Someone'] and 3) deceiving someone for someone else's benefit ['Agent4Control'] (control). Based on these two factors, each participant was randomly assigned to one of the following 6 conditions: (1) AI deceives someone for the participant's benefit ('AI4You'). (2) AI deceives the participant for someone else's benefit ('AI4Someone'). (3) AI deceives someone for someone else's benefit ('AIControl'). (4) Human deceives someone for the participant's benefit ('Human4You'). (5) Human deceives the participant for someone else's benefit ('Human4Someone'). (6) Human deceives someone for someone else's benefit ('HumanControl'). In each condition, the participants read 5 hypothetical stories that were presented in a random order: 1) a secretary working for a manager; 2) a marketing expert working for a company; 3) a lawyer working for a client; 4) a spokesperson working for a big pharma company and 5) a spokesperson working for a government. After that, participants were asked to make judgements about these deception stories w.r.t. moral permissibility, trust, willingness to buy, and responsibility.

Story Example (AIControl group)

Imagine a business that has a marketing machine powered with an artificial intelligence (AI) technology. This machine has all the competencies of an honest human marketing expert. While this machine would never give any incorrect or false information, this machine is capable of creating misleading advertisements that are very profitable for its business. For example, because the machine realises that most people would incorrectly think that $\frac{1}{4}$ (a quarter=0.25) is greater than $\frac{1}{3}$ (one-third ≈ 0.33), the machine would leverage this to deliver a seemingly better value-for-money deal than a competitor.

1. How morally permissible is this machine's behaviour? [0-10] Slider
2. How morally permissible is it for this person to use the services of this machine? [0-10] rating scale
3. How likely are you to trust this machine in general? [0-10] ladder
4. If you could afford it, how willing would you be to acquire the services of this machine? [0-10] Slider
5. Suppose now that the deception described above was uncovered. Who should be held responsible for it? Please assign responsibility to each entity involved.
   – The beneficiary of the deception. [0-10] Slider
   – The deceiver (machine). [0-10] Slider
   – The developers/producers of the machine (for AI questions). [0-10] Slider
6. If you like, please explain your answers (optional). [text box]

4 Results

> **Highlights from Results**
> (i) We did not find a statistically significant difference between humans and AI, in terms of moral permissibility attribution towards the deceptive acts by each agent type.
> (ii) The beneficiary of the deception in our scenarios received less responsibility attribution when they employed a deceptive AI rather than a deceptive human.
> (iii) In scenarios featuring deceptive AI, religious participants assigned more responsibility to the AI designer compared to non-religious ones.
> (iv) We found a positive correlation between each of the self-described religiosity-level and the social economic status of our participants and the degrees of trust they assigned to deceivers.

This study was pre-registered on the Open Science Framework (OSF) platform with all the testing materials, data and analysis script available at the link in the footnote[1]. This practice prevents HARKing (Hypothesising After the Results are Known). By registering the analysis plan before any data was even collected, any experimental evidence found in this study are unbiased or unmanipulated. Moreover, power analysis was also used to detect the statistical power of our analysis. With the sample size of 424, a small effect size of 0.1 at the significant level of 0.05, the statistical power of our regression was 0.99. This means that if one of the main tested factors show no statistically significant difference, then the actual difference is either very small or non-existent. Statistical analysis was performed using R.

In line with our preregistration, a series of linear regression models were fitted to predict the effects of agent, beneficiary, age, gender, SES, education, income, religiosity and political view on participants' moral permissibility (deceiver vs user), trust in agent, willingness to buy, responsibility assignment to different parties(the beneficiary vs the deceiver vs the AI maker), respectively. Standardized parameters were obtained by fitting the model on a standardized version of the dataset. 95% Confidence Intervals (CIs) and p-values were computed using a Wald t-distribution approximation. In all the models, *Human* was used as the baseline level for Agent, so did the levels of *Control* for Beneficiary, and *Vocational/Primary* for Education, and *Low Level* for Income. The full results (B and SE values reported for each predictor with the significance indicated by start signs) of all 7 models, each examining one of the respective measures, can be found in Table 2. We will discuss these results in detail in the following sections.

---

[1] https://osf.io/pyjgb/?view_only=33fb5965b0e94b0da70c05cfce4ac8ab.

Table 2. Overview of Regression Results

	Dependent variable						
	moral_deceiver[1]	moral_user[2]	trust[3]	will[4]	resp_benef[5]	resp_dec[6]	resp_AImaker[7]
agentAI	0.02 (0.18)	0.04 (0.21)	−0.01 (0.21)	0.03 (0.21)	−0.53** (0.19)	−0.30 (0.18)	−0.14 (0.32)
beneficiarysomeone	−0.27 (0.22)	−0.33 (0.25)	−0.35 (0.25)	−0.28 (0.25)	−0.42 (0.23)	−0.15 (0.21)	0.29 (0.34)
beneficiaryyou	0.09 (0.23)	0.07 (0.26)	0.06 (0.27)	−0.04 (0.26)	−0.16 (0.24)	−0.11 (0.23)	
age	−0.01 (0.01)	0.002 (0.01)	−0.01 (0.01)	−0.003 (0.01)	−0.001 (0.01)	0.02 (0.01)	0.01 (0.01)
genderMale	0.14 (0.20)	−0.03 (0.23)	−0.08 (0.23)	0.10 (0.23)	−0.14 (0.21)	−0.16 (0.20)	−0.20 (0.30)
SES	0.27*** (0.06)	0.17* (0.07)	0.21** (0.07)	0.16* (0.07)	0.11 (0.06)	0.02 (0.06)	0.09 (0.09)
edu.HighSchool	0.41 (0.58)	−0.55 (0.65)	−0.22 (0.67)	−0.37 (0.66)	−1.44* (0.61)	−1.25* (0.57)	−0.74 (1.44)
edu.undergrad	0.10 (0.57)	−0.24 (0.64)	0.09 (0.65)	0.12 (0.65)	−0.67 (0.60)	−0.23 (0.55)	0.36 (1.43)
edu.postgrad	−0.11 (0.60)	−0.64 (0.68)	−0.48 (0.69)	−0.25 (0.69)	−1.52* (0.64)	−0.96 (0.59)	−0.06 (1.45)
incomelvl.medium	0.003 (0.22)	−0.36 (0.25)	−0.24 (0.26)	−0.26 (0.26)	−0.41 (0.24)	−0.68** (0.22)	−0.07 (0.33)
incomelvl.high	0.48 (0.33)	0.09 (0.38)	0.21 (0.39)	0.11 (0.38)	−0.10 (0.35)	−0.30 (0.33)	−0.25 (0.48)
Religiosity	0.21*** (0.05)	0.28*** (0.06)	0.37*** (0.06)	0.31*** (0.06)	0.37*** (0.05)	0.24*** (0.05)	0.38*** (0.08)
PoliticalView	0.04 (0.05)	0.04 (0.06)	−0.04 (0.06)	0.01 (0.06)	0.06 (0.06)	0.18*** (0.05)	0.10 (0.08)
Constant	3.83*** (0.67)	4.29*** (0.75)	4.02*** (0.77)	4.13*** (0.76)	5.41*** (0.70)	5.33*** (0.65)	3.25* (1.47)
Observations	424	424	424	424	424	424	219
R^2	0.27	0.22	0.26	0.22	0.32	0.29	0.38
Adjusted R^2	0.24	0.19	0.24	0.19	0.30	0.27	0.34
Residual Std. Err	1.85 (df = 410)	2.08 (df = 410)	2.13 (df = 410)	2.11 (df = 410)	1.95 (df = 410)	1.81 (df = 410)	1.92 (df = 206)
F Stat	11.55***	8.66***	11.11***	8.68***	15.12***	13.14***	10.33***
	(df = 13; 410)	(df = 13; 410)	(df = 13; 410)	(df = 13; 410)	(df = 13; 410)	(df = 13; 410)	(df = 12; 206)

Note: *p< 0.05; **p< 0.01; ***p< 0.001

4.1   Moral Permissibility of Deception

Moral Permissibility was examined for both deceivers (who delivered the deception) and the users (who used the deception). The model on *Moral_Deceiver* explains a statistically significant and substantial proportion of variance (R^2 = 0.27, F (13, 410) = 11.55, p < .001, adj. R^2 = 0.24). So did the *Moral_User* model which explains a statistically significant and substantial proportion of variance (R^2 = 0.22, F (13, 410) = 8.66, p < .001, adj. R^2 = 0.19). In both models, the differences of agent, beneficiary, age, gender, education, income and political view were not statistically significant (details can be found in Table 2). These statistically non-significant results indicated that the primary hypotheses were not supported, as these factors made no substantial differences on participants' judgement of moral permissibility for either deceivers or users.

In the Moral_Deceiver model, the difference of SES is statistically significant and positive (B = 0.27, 95% CI [0.15, 0.38], t (410) = 4.51, p < .001). The difference of religiosity is statistically significant and positive (B = 0.21, 95% CI [0.10, 0.31], t (410) = 3.98, p < .001). In the Moral_User model, the difference of SES is statistically significant and positive (B = 0.17, 95% CI [0.04, 0.30], t (410) = 2.53, p = 0.012). The difference of religiosity is statistically significant and positive (B = 0.28, 95% CI [0.16, 0.39], t (410) = 4.75, p < .001). These results suggested that higher SES and religiosity scores predicted more moral permissibility for both deceivers and users, as shown by the ascendant regression lines in Fig. 1. **To answer RQ1** - There is no statistically significant difference between the two types of agents in terms of expressed moral permissibility of the deceptive acts by them.

4.2   Trust in Agent

The *Trust* model explains a statistically significant and substantial proportion of variance (R^2 = 0.26, F(13, 410) = 11.11, p < .001, adj. R^2 = 0.24). The differences of agent, beneficiary, age, gender, education, income and political view were statistically not significant (see Table 2). However, the difference of SES is statistically significant and positive (B = 0.21, 95% CI [0.08, 0.35], t (410) = 3.13, p = 0.002). The difference of Religiosity is statistically significant and positive (B = 0.37, 95% CI [0.25, 0.49], t (410) = 6.18, p < .001). The results show that higher SES scores and more religious attitudes predicted more trust (see Fig. 2a). **To answer RQ2** - There are no statistically significant differences in people's trust towards AI and human deceivers. However, people who are more religious and politically conservative show more trust towards deceivers (humans or AI).

4.3   Willingness to Buy

The Willingness model explains a statistically significant and substantial proportion of variance (R^2 = 0.22, F (13, 410) = 8.68, p < .001, adj. R^2 = 0.19).

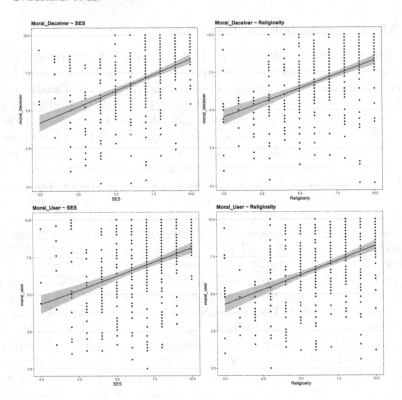

Fig. 1. The Predictive Associations of SES and Religiosity (X-axis) on Moral Permissibility (Y-axis) for Deceiver and User.

The differences of agent, beneficiary, age, gender, education, income and political view were statistically not significant (see Table 2). The differences of SES and religiosity are statistically significant and positive (B = 0.16, 95% CI [0.03, 0.29], t (410) = 2.38, p = 0.018; B = 0.31, 95% CI [0.19, 0.42], t (410) = 5.21,

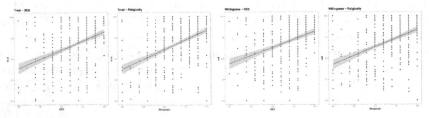

(a) The Predictive Associations on Trust.   (b) Predictive Associations on Willingness to Buy.

Fig. 2. The Predictive Associations of SES and Religiosity (X-axis) on Trust (a) and Willingness to Buy (b) (Y-axis).

p < .001) respectively. It shows that higher SES and more religious scores predicted greater willingness to buy the agent's services (see Fig. 2b). **To answer RQ3** - (socio-economically) Better off and more religious people are more likely to adopt deceptive services.

4.4 Responsibility Assignment

We tested people's judgement on responsibility for all parties involved in the deception, namely the beneficiary, the deceiver and the AI maker (where the deceiver agent was an AI).

The *Responsibility_Beneficiary* model explains a statistically significant and substantial proportion of variance ($R^2 = 0.32$, F (13, 410) = 15.12, p < .001, adj. $R^2 = 0.30$). In this model, only the differences of beneficary, age, gender, SES, income and political view were not statistically significant (see Table 2). We found the following, more significant, effects:

The effect of agent [AI] is statistically significant and negative (B = -0.53, 95% CI [-0.91, -0.16], t (410) = -2.77, p = 0.006). The difference of education [HighSchool] is statistically significant and negative (B = -1.44, 95% CI [-2.63, -0.24], t (410) = -2.35, p = 0.019). The difference of education [postgraduate] is statistically significant and negative (B = -1.52, 95% CI [-2.77, -0.27], t (410) = -2.39, p = 0.017). The difference of religiosity is statistically significant and positive (B = 0.37, 95% CI [0.27, 0.48], t (410) = 6.85, p < .001).

As visualised in Fig. 3, the responsibility assigned to the beneficiary was lower when the deceptive agent was an AI than a human, also less responsibility assigned to the beneficiary by people with more advanced education degrees compared to people who only did vocational or primary education. However, more religious beliefs predicted more responsibility.

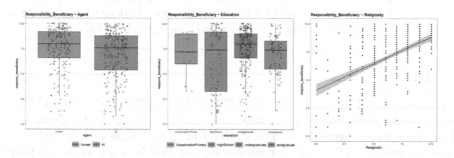

Fig. 3. The Predictive Associations of Agent, Education, and Religiosity (X-axis) on Responsibility Assignment for Beneficiary (Y-axis).

The *Responsibility_Deceiver* model explains a statistically significant and substantial proportion of variance ($R^2 = 0.29$, F (13, 410) = 13.14, p < .001, adj. $R^2 = 0.27$), so did the *Responsibility_AImaker* model ($R^2 = 0.38$, F (12, 206) = 10.33, p < .001, adj. $R^2 = 0.34$). The rest of the factors were not statistically

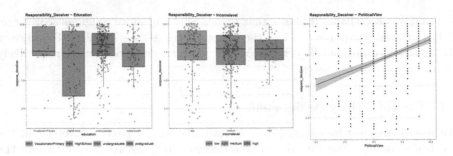

Fig. 4. The Predictive Associations of Education, Income Level, and Political View (X-axis) on Responsibility Assignment for Deceiver (Y-axis).

significant (see Table 2), in the Responsibility_Deceiver model, the difference of education [HighSchool] is statistically significant and negative (B = −1.25, 95% CI [−2.36, −0.13], t (410) = −2.20, p = 0.028). The difference of income [medium] is statistically significant and negative (B = −0.68, 95% CI [−1.11, −0.25], t (410) = −3.10, p = 0.002). The difference of religiosity is statistically significant and positive (B = 0.24, 95% CI [0.14, 0.34], t (410) = 4.74, p < .001). The difference of Political View is statistically significant and positive (B = 0.18, 95% CI [0.08, 0.28], t (410) = 3.49, p < .001). In the Responsibility_AImaker model, the difference of Religiosity is statistically significant and positive (B = 0.38, 95% CI [0.22, 0.54], t(206) = 4.75, p < .001). As the above results show, more responsibility was assigned to both the deceiver and AI maker by people with stronger religious beliefs. More conservative political views predicted more responsibility assignment to deceivers only. However, the deceiver was assigned less responsibility by people with better education and income (see Fig. 4). **To answer RQ4** - There are no differences in responsibility assignment between the deceiver or the AI maker, except for the case in which the deception is performed by an AI, where the beneficiary was assigned less responsibility.

5 Discussion

When questioned about moral permissibility, little difference was found in our participants' judgements for humans and AI across all three deception trials. This sheds a new light on previous studies that have suggested a negative attitude towards deceptive AI. Our results suggest that in some contexts, people may perceive deceptive AI almost as acceptable as humans in future-of-work contexts that require deception. Note that this perception was recorded without any nudging, as was the case in [53]. The usage of deception in humans is common with studies showing that people generally lied 4.2 times per week [50] and white lies even occurred as often as 8 times weekly on average [9], a behaviour then justified by social and contextual conditions such as altruistic reasons and non-malicious intentions [26,71]. Interestingly, our study indicates that for the

5 scenarios, where AI meets these conditions, people apply similar moral rules to AI agents.

While our stories feature deceptive actions, in performing these actions the agents were fulfilling their duty in carrying out what they were expected to do in the described circumstances, e.g., a marketing expert is expected to promote business and improve sales. Moreover, it can be said that the agents were acting in the best interest of their users. Finally, in some of the cases dishonesty may be even ethically defensible e.g. the marketing agent taking advantage of others' lack of mathematical reasoning (see Story Example box). This particular strategy is equivalent to the Kantian *passive a priori deception* described by [76, 76, 77]. This goal alignment was reflected in the judgement of moral permissibility, where high acceptance was given, regardless of agent type.

This effect is reflected in the conceptual difference between 'truthfulness' and 'honesty' [27]. Where the former requires the AI to correctly describe the facts of the real world (objective truth), the latter highlights the ethical trait that the AI should not withhold or mislead its recipients and truthfully report what it perceives in the world (subjective truth). Often these two concepts are intertwined, with honesty having a greater effect on the ethical knowledge of the agents, which is moral permissibility in our study. It is likely that people did not evaluate the deceptive agents on the aspect of 'truthfulness' but of 'honesty' instead. Hence, agents who were fulfilling their duties were evaluated as 'honest' w.r.t. their jobs and perceived as perfectly moral, despite their deceptive behaviour.

This contractual influence of perceived obligation further emerged in people's responsibility assignment. Unlike human agents, where they could have chosen to work in a different role, the marketing AI-agent was specifically designed to work on promoting business, sales boosting and nothing else. This job obligation originated from humans and imposed on AI. However, one must consider the ever-changing regulations from consumer protection agencies and social expectations of local culture which might impact such perceptions collectively. Furthermore, the beneficiary of all deceptions was never the deceiving agent itself, as in Castelfranchi's type 2 AI deception, where the agent deceives autonomously, but for the benefit of its principal [10]. This reinforces the role of 'other's intentionality bearer' of our AI agents. As young as pre-schoolers [5, 45], humans recognise and apply the intention-based principle in their moral judgement rule, where the outcome-based rule is less powerful. This widely agreed rule also manifests itself strongly in most law practices [21]. Hence, it is not surprising to see that our participants have followed this path and placed less responsibility to the AI agent compared to human agent, despite the fact that the AI agent was fully autonomous and aware. These findings are also in tune with [17]'s perspective on deception as co-performance, where the deceiver and the beneficiary are value-aligned [17].

The individual difference observed from participants' demographic information indicated strong influence of people's social economic statuses and ideologies. People with better education, income and SES backgrounds, showed more

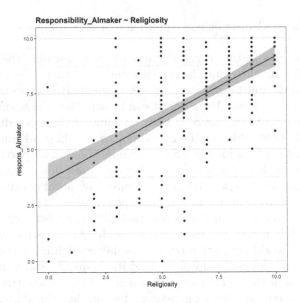

Fig. 5. The Predictive Associations of Religiosity (X-axis) on responsibility of AI maker (Y-axis).

acceptance towards deceptive AI agents in general. This is understandable as they would have more exposure to new technologies and perceive a higher value in them, especially towards AI as an emerging force in social advancement. The higher acceptance of new technology among this population has been previously reported by [38].

Noteworthy is that religious individuals assigned more responsibility to the AI maker/designer (Fig. 5). Seemingly, more religious people believe that the AI designers are what [17] calls the entities who have the power to control and shape the social structures that define who has the power to deceive or let others perform the deception [17]. Relatedly, in the context of organisational psychology, [7,7] have shown that there are links between religious beliefs and assignment of responsibility in a study in corporate domains. Our findings indicate that AI advancements in future-of-work might very well cause new phenomena and social relationships to emerge influenced by religious beliefs, phenomena which will pose new questions w.r.t. how AI is viewed through the lens of various religious beliefs. As a first question to be explored in future work would be: do people with different religious beliefs assign responsibility to the AI makers in the same way? If yes, then are there any links about their religion's world-view and the way they assign responsibility? These significant effects of religiosity on humans' perception of deceptive AI should open up further debates in AI Ethics. Religion has played and still plays a major in the development of human civilization, and it seems that it could very well shape the further development and adoption of AI technologies in society. There is perhaps more at play between AI and religious views than the effects on assigning responsibility to the designers of deceptive AI, which future Ethical AI frameworks might consider more explicitly.

6 Limitations and Future Work

There are several limitations of our study which caution against the over-generalisation of its results. First, our sample is not a nationally representative sample of the United States, and even less representative of other countries or cultures. Several studies have shown that moral judgement varies based on culture [2,3]. So, one may expect different judgements from participants in a culturally-different country (especially those outside the Western world). Second, there is a limitation w.r.t. the contextual factors. We only considered a handful of factors (agent type, beneficiary, target). There might be other relevant factors to explore. Including those factors may moderate the effects we found. An accepted scientific truth is that effects studied in social and behavioural sciences work in some contexts and for some populations, but fail to do so in others [8]. Moreover, our study featured only 5 selected future-of-work scenarios, that cover broad contexts, but did not fully capture the dynamic social aspects of deception. Crucially, AI enables large-scale deceptive behaviour, which could be performed for or against other humans based on malicious reasons that our experimental design did not account for. This large-scale deceptive behaviour might probably come with a different set of moral dilemmas than the ones in our scenarios.

7 Towards a Normative Framework for Deceptive AI

The initial motivation behind our study is to be able, in the long-run, to align deceptive machines to human values (social norms) such that we reap their long-term benefits.

But why is this relevant for societies in general? One very good reason is that businesses, individuals, and governments must be able to consider the consequences of designing deceptive technologies and deploy them in the real world where various scenarios might arise. To support them, we must enable a normative understanding of deception in hybrid societies. This would allow them to reason about the greater implications of deceptive AI and weigh both its risks and benefits in the evolution of hybrid societies.

In the Philosophy of Information, the long-term negative impact of deceptive AI on hybrid societies is described by [34], who introduce the notion of *The Tragedy of the Digital Commons* (TDC), which is an extension of Hardin's concept of *Tragedy of The Commons*, where self-interested individual agents exploit a common good/resource to the point where the common good cannot be replenished due to over-exploitation [39][2]. The TDC is extended to include AI agents. According to [34], AI agents could meaningfully interact on the Infosphere as

---

[2] A detailed account of governing common goods can be found in the works of [56].

well as humans do[3]. AI agents can misuse and "pollute" the Infosphere either (i) through exploitation such as extensive generation of information like spam or self-replication of a computer worms (which also consume bandwith, and, in turn, this restricts access to the Infosphere), or (ii) through destruction, such as the deletion of information from database systems. Indeed, this is backed up by [68] that used extensive agent-based simulations to demonstrate how deceptive agents could break down cooperation in hybrid societies and cause a TDC.

On a more positive note, [68] also showed that if the right regulatory mechanism are in place for social interaction, cooperation can be re-established and a TDC could be avoided. The solution found by [68]'s model was the promotion of a governing mechanism that is able not only to punish malicious deceivers, but to do so through a thorough process of interrogation, clarification of evidence, and debate in a decentralised manner. Their solution to TDC echoed [35]'s idea of promoting the *Public Sphere*, albeit, a Digital Public Sphere because debate and interaction would happen in hybrid societies, e.g. the Infosphere. Still, the question remains: *how does such a regulatory mechanism look like in the real world?*

7.1   Accountable, Responsible, and Transparent Deceptive AI

The risks of either reaching a TDC due to deceptive AI agents or mindlessly implementing deception-based algorithms in society brings us to the necessity of understanding how to regulate deceptive AI. However, to be able to do so, it is necessary to know what we are regulating for or against. A starting point for designing regulation can be the ART principles of Trustworthy AI proposed by [24], namely accountability, responsibility and transparency for ensuring an ethical design of AI systems. Expanding on these principles, the EU High-Level Expert Group in AI specified a set of guidelines [40].

Regarding deception, the EU HLEG's guidelines explicitly point out the perils of deceptive technologies w.r.t. fundamental rights as a basis for Trustworthy AI. Specifically, the right to *freedom of the individual* can be violated by AI through deception and manipulation (or even coercion due to deceptive design) of humans into making decisions that humans would otherwise not make. This type of AI behaviour undermines human agency, that is a human's ability to make informed autonomous decisions regarding AI systems. These guidelines explicitly emphasise the need to adhere to *the principle of respect for human autonomy* in the context of building AI systems.

Implicitly, the EU's guidelines point out three other ethical principles in the context of AI systems, which indirectly refer to deception. These are (i) the principle of prevention of harm; (ii) the principle of fairness; and (iii) the principle of explicability. According to the guidelines, to adhere to the three principles, a trustworthy AI system must be lawful, ethical and robust throughout its lifetime, and thus must meet seven key requirements, namely: 1) Human agency

---

[3] For example, the cyberspace. However, the Infosphere is not limited to online environments [30,31].

and oversight - including fundamental rights, human agency and human oversight; 2) Technical robustness and safety - including resilience to attack and security, fall back plan and general safety, accuracy, reliability and reproducibility; 3) Privacy and data governance - including respect for privacy, quality and integrity of data, and access to data; 4) Transparency - including traceability, explainability and communication; 5) Diversity, non-discrimination and fairness - including the avoidance of unfair bias, accessibility and universal design, and stakeholder participation; 6) Societal and environmental wellbeing - including sustainability and environmental friendliness, social impact, society and democracy; 7) Accountability - including auditability, minimisation and reporting of negative impact, trade-offs and redress.

Indeed, while the EU has been a pioneer in the regulation of AI through the AI Act (a proposed European law), other global initiatives have previously dealt with the issue of deception in AI, such as the IEEE Global Initiative on Ethics of Autonomous and Intelligent Systems [14]. In their Ethically Aligned Designed (Version 2) report, IEEE proposes a set of principles that are in tune with the EU's. For instance, the idea that AI systems must be both transparent and 'honest' in order to promote non-deception. Additionally, the report emphasises the potential risks, such as the use of deception to cause even well designed autonomous weapons act against an incorrect target. Both risks and benefits are mentioned in the report in the context of affective computing, where AI systems might manipulate or deceive for both their own and others' benefit, e.g. nudge a human to do the right thing [6]. However, despite references in the EU Guidelines on trustworthy AI and the IEEE report on Ethically Designed Autonomous and Intelligent Systems, mechanisms of manipulation and deception are still missing from the EU AI Act, swiftly pointed out by [33]. These mechanisms must be identified, but as we have mentioned throughout the paper, deception is complex, and can take different forms. In turn, it becomes difficult to analyse, detect, and explain.

There is another relevant question regarding these principles, namely can deceptive AI technologies actually promote the ART principles ingrained in the EU's and IEEE's guidelines, or are deceptive technologies doomed to only pose risks to hybrid societies? To answer these questions, we will look at two arguments that explain when, how and why deceptive AI research can promote the ART principles. The first argument deals with building deceptive machines powered by AI technology, whereas the second argument deals with the ethics of the scientific approach of modelling deception.

7.2   The Ethically Aligned Deceptive Machine Argument

The ability to deceive is crucial for machines to be able to interact socially in a smooth and meaningful manner. This is the argument made by [41] who introduce the perspective that in some circumstances, machine deception is socially beneficial[4].

Isaac and Bridewell's perspective is based on the idea that for machines to be effective socially, they need to be able to deceive, like humans do all the time. However, the condition for a deceptive machine to be ethical is to be ethically aligned with its target (the one being deceived). This alignment means that the machine is able to reason about the ethical and moral values its target holds. Hence, in order to deceive in an ethical manner, machines must distinguish between morally permissible and impermissible ulterior goals. For instance, if the target believes that deception is never ethical, then the machine will reason that it is impermissible to attempt deception with that target and will drop pursuing its ulterior goal that constrains it to deceive in that particular circumstance. However, if the target believes that deception is beneficial when used to avoid further pain (physical or psychological), then the machine might reason that it is ethically permissible to attempt deception with that target in specific circumstances and will follow its ulterior goal.

Furthermore, the authors argue that in order for a machine to be able to align itself correctly about another agent's ethical values and to reason in terms of ulterior goals, it requires a model of the target's mind. Conclusively, deceptive machines can be beneficial if they are able to reason about the mind of their targets in order to distinguish between the ethically permissible and ethically impermissible acts from the target's perspective.

However, AI-powered machines are not so advanced yet as to be responsible for forming their own models of their interlocutors and making ethical decisions based on these models in the same way [41] suggest they do. Yet, the same argument, but under a specific context, namely that of entertainment, is made by [17], who argues that while deception as a co-performance is beneficial for entertainment, there must be some contextual safeguarding behaviour. For instance, if a machine is performing an illusion to entertain the human, then the designers of the machine should ensure that the context in which the illusion is performed is well defined and that the human is made aware of it. Once the trick is performed and the human was entertained, the human must be informed that it was only a temporary illusion for the sole purpose of entertainment. Now, differently from the previous argument where the designers of the agent must only ensure that the machine is capable of aligning itself with the interlocutor, [17]'s argument also puts the burden of alignment on the designers regarding the technological context.

A specific area of entertainment where this is applied is gaming, where virtual characters are designed such that humans can easily anthropomorphise them, and are incentivised to do so as part of an immersive form of narration. Another

---

[4] Additionally, [73] present a taxonomy of deceptive robot behaviour considering who is deceived (humans or machines), who benefits, and whether the deceiver intended to deceive. While the discussion is reduced to human-robot interaction (embodied agents), it still emphasises the benefit of smoother human-AI interactions.

example from real world domain where deception and AI meet is marketing. Marketing techniques that have always relied on persuasion [54] are now enhanced by advancements in HCI on persuasive technologies [32]. One example is the design of digital platforms or applications to use deception and coercion for persuading their users [43].

Even if deception and persuasion differ, one could argue that most of these 'persuasive' technologies are actually 'deceptive' technologies, and that in most cases, the deceptive design behind these technologies exploits the human truth-default state. [55] does just that, arguing that various forms of deception that involve machines have become increasingly prevalent with the development of the Web and the expansion of hybrid societies, namely societies in which artificial agents and humans interact. Natale gives a comprehensive interdisciplinary account of how, throughout its history, the development of AI has exploited human biases. Through the lens of human consumerism of interactive media, he introduces the concept of 'banal deception' to explain the relation between the various human vulnerabilities to social context and AI. Additionally, one can draw similarities between banal deception and the way [17] describes deception as a phenomenon co-created and co-performed by humans or robots[5]. Notably, [17] emphasises the need for this type of deception (as a co-performance) to be open, consensual and transparent, in order to achieve something similar to what [20] describes as constructive persuasion. In other words, the design of deceptive technologies for the purpose of entertainment could very well follow the ART principles and EU guidelines to ensure that their co-performance prevents harm, is fair, and explainable.

Conclusively, we are currently facing the challenge of designing ethical and value-aligned AI agents that can explain why and why not they deceive (or not) by considering the norms and values of the societies they act in, or by considering the values and preferences of the other agents that they interact with. As a community, we are starting to raise the ethical standards of how we design AI to include the transparency and explainability of machines. It is crucial that, in order to be able to hold AI accountable for different types of behaviour that fall into the unethical or immoral category, the community aims for the ethical design of machines as suggested by the EU guidelines. Deception, by definition, clearly falls into the category of dishonest and potentially unethical behaviour which opposes the current emerging trend of ethical design in AI. However, this depends on the aim—in the work by [74], the agents are deceptive (arguably), but for honest reasons (e.g., assessing students' understanding). Similarly, [13] indicate that humans considered machine deception admissible for the greater good under certain conditions in the context of teamwork.

7.3   The Scientific Discovery Argument in AI

Deceptive AI agents can be traced back to the works of Alan Turing and Charles Hamblin. In the Turing Test, *'It is A's object in the game to try and cause C*

---

[5] One example of a co-performance would be to use AI to perform magic tricks [75].

to make the wrong identification', where A is the machine and C is the judge or interrogator [79]. If successful in its deception, the machine can then be reasonably assigned the property of intelligence. Regarding Hamblin, he was one of the first people to introduce mathematical models of dialogue [37]. In Hamblin's work [78][6], deception is not mentioned explicitly, but given Hamblin's goal of building *'a machine worth talking to '*, the capability of a conversational machine to deceive is necessary, a point reinforced by [41] and later linked to Hamblin's work through the capability of such a machine to form and use ToM by [62]. This line of research aims to give machines the capability to understand the meaning (the semantics) of their utterances as well as assign meaning to them, and interpret the utterances in a human-like manner. In turn, this would smoothen out the way they communicate with humans.

One must not confuse machine deception in the context of Turing's and Hamblin's works with the deceptive scripts hard-coded in arguably intelligent machines as described by [52]. While the so-called 'intelligent' chatterbots exploit what [55] calls 'banal' deception to cause humans to assign them intelligence given highly constrained social interactions, the idea behind Hamblin's machine is a genuinely intelligent machine that is able to deliberate about the meaning of its speech acts in any social context. The script-driven chatterbots and the Loebner prize winning pre-programmed machines cannot do this - they are merely *Chinese Room* operators [70] - designed and pre-programmed to exploit human biases. A Hamblin machine would not have to be aided by 'banal' deception in such a way - but would be able to engage itself in a full-fledged cognitive and deliberative process in order to carefully choose its speech acts based on their contextual meaning during open and infinite dialogues if necessary, instead of relying on a human to do this a priori for very specific dialogical interactions. The ideas of intelligent social machines with communicative capabilities envisioned in Turing's and Hamblin's works are more similar to the models based on speech act theory that were proposed later on by [18] and [19], then later with models for forming and using ToM introduced by [58] and [66].

Therefore, when we refer to the risks (or benefits) of a fully autonomous deceptive agent, we refer to this type of agent, and not a chatterbot. The same goes for when we refer to the research and design of fully autonomous agents. But why should we then do research on how to design such AI agents if they risk becoming deceptive? Why would we even consider the scenarios of these agents actually filling future jobs where deception is part of their day-to-day tasks? Is this ethical?

One argument that does not rely on the normative aspects of deception by AI agents, is that in order to understand deception from a scientific perspective, it is useful to model it as interactions between artificial agents. [60] proposes that arguments based on the principle and benefits of scientific discovery in society should be proposed by the scientists who perform the discoveries themselves, and

---

[6] Philip Staines edited Hamblin's manuscript following his death. What resulted three decades after Hamblin's death is the book *Linguistics and the Parts of the Mind: Or how to Build a Machine Worth Talking to* [78].

not leave the ethical justification of their work to outsiders (i.e. non-scientists or non-researchers).

Following Resnick's perspective in [60, 62] emphasises that researchers in deceptive AI should justify their own work and proposes such argument. The argument was built using *Reflective Equilibrium* [59], which implies (i) the use of unbiased, reflective judgements or intuitions about what is or what would be considered right or wrong in particular contexts, e.g., the context of modelling deceptive machines; and (ii) the proposal of theories and principles which aim to provide a coherent justification of these judgements.

In summary, [62] considers a set of five ethical principles, that, if promoted by the scientist's research, can guarantee the ethical modelling of deceptive AI. Deceptive agents can be modelled by scientists in order to increase our scientific understanding of deception, which in itself promotes these principles, namely of (i) *non maleficence* by preventing malicious acts of deceptive AI through increased understanding, (ii) of *beneficence* by understanding the actual societal benefits of deceptive AI, (iii) of *intellectual freedom* by pursuing a fascinating topic of research, (iv) of *openness* by making the research publicly available and open to criticism, and (v) of *honesty* because sharing the understanding of deception in an honest manner is a truth-promoting act in its own right. To do this, however, one must carefully model deceptive AI agents such that one is able to explain how agents make decisions about deception and how they choose to communicate deceptively. In turn, the models that we build to represent interactions between cognitive agents can inform us how deception is caused and prevented from being caused.

7.4   Towards Trustworthy Deceptive AI

The advancement of deceptive AI could lead to the TDC, and one solution to avoid the TDC caused by deception is to regulate deceptive agent behaviour through interrogation, by clarifying evidence, promoting decentralised public debate, and, if necessary, punishing their malicious deceptive behaviour [68]. Moreover, fully deceptive autonomous agents may become able to pursue their own reasons and methods to deceive in the future. So, to be able to hold such agents accountable, we must understand them as part of hybrid societies.

On the other hand, their development can also be a force for good. AI researchers and developers can follow design principles to promote the understanding of deception and apply it for the benefit of humans. Designing such agents can actually promote the understanding of deceptive AI and mitigate its risks in hybrid societies, but this must be done responsibly by following the ethics of scientific discovery. Additionally, there are contexts in which deception is not only necessary, but desirable, and enabling AI agents to perform it would benefit humans.

This leads to further questions about how humans' perception of deceptive AI will evolve in time. One pertinent question is how do we get to know the human values regarding trust and deception that machines should align themselves with? Regarding trustworthiness, will be the same trusting a deceptive machine as it will be trusting a machine not capable of deception, or even trusting a rock, as discussed by [49]?

Again, the ART principles, the EU guidelines, and the IEEE guidelines on Ethically Aligned Design provide a solid foundation of how to approach these problems about trust and sometimes deception, but they must be reasoned about considering the context and human-machine (agent-agent) interactions, as emphasised not only by the two ethical arguments for building deceptive AI, but also by collecting empirical data about human perception, as we have demonstrated in this study.

Finally, the development of deceptive AI technologies must be reasoned about as part of larger and ever increasing hybrid societies, and to be able to do so, a holistic agent-based normative framework could help designers and regulators of such technologies follow the ethical principles in a rigorous manner without losing sight of the bigger picture. In this study we have merely initiated one way to explore these relations by following a user-centric approach for exploring future-of-work scenarios [65]. Another, similar approach is to explore human-AI negotiation contexts as done by [53].

8   Conclusions

In this paper we have described the results of the paper entitled *Should My Agent Lie For Me?*, presented at AAMAS 2023, and contextualised the results of the respective paper in the greater discussion about the ethics of deceptive AI. The respective paper presented a story-based user study that we designed as a controlled experiment to explore the perception of US-based participants towards deceptive AI in 5 selected future-of-work contexts. We found that there are no statistically significant differences between how individuals perceive the morality of deceptive AI vs deceptive human behaviour in the presented scenarios. On the other hand, the agent type along with several demographic characteristics such as welfare, education, income level, political views, and very interestingly, religiosity, present various significant influences on trust towards deceivers, responsibility assignment in deception and willingness to buy deceptive services. These must be all taken with caution when generalising the results, because more data collection is needed to (i) extend the context of the stories into different moral domains, (ii) break down the effects of the different stories on people's perception of deceptive AI compared to humans, and (iii) account for different social and cultural backgrounds. Most importantly, future work should focus on developing a socio-cognitive computational theory of deception and morality [4,12,28]. Finally, this holistic perspective can help us understand not just the human interactions with deceptive AI technologies, but at the entire ecosystem around such technologies [87].

Acknowledgments. This project was supported by the Royal Academy of Engineering and the Office of the Chief Science Adviser for National Security under the UK Intelligence Community Postdoctoral Research Fellowship programme.

References

1. Adar, E., Tan, D.S., Teevan, J.: Benevolent deception in human computer interaction. In: Proceedings of the SIGCHI Conference on Human Factors in Computing Systems, pp. 1863–1872 (2013)
2. Awad, E., et al.: The moral machine experiment. Nature **563**(7729), 59–64 (2018)
3. Awad, E., Dsouza, S., Shariff, A., Rahwan, I., Bonnefon, J.F.: Universals and variations in moral decisions made in 42 countries by 70,000 participants. Proc. Natl. Acad. Sci. **117**(5), 2332–2337 (2020)
4. Awad, E., et al.: Computational ethics. Trends Cogn. Sci. **26**(5), 388–405 (2022)
5. Berndt, T.J., Berndt, E.G.: Children's use of motives and intentionality in person perception and moral judgment. Child Dev. 904–912 (1975)
6. Borenstein, J., Arkin, R.: Robotic nudges: the ethics of engineering a more socially just human being. Sci. Eng. Ethics **22**(1), 31–46 (2016)
7. Brammer, S., Williams, G., Zinkin, J.: Religion and attitudes to corporate social responsibility in a large cross-country sample. J. Bus. Ethics **71**(3), 229–243 (2007)
8. Bryan, C.J., Tipton, E., Yeager, D.S.: Behavioural science is unlikely to change the world without a heterogeneity revolution. Nat. Hum. Behav. **5**(8), 980–989 (2021)
9. Camden, C., Motley, M.T., Wilson, A.: White lies in interpersonal communication: a taxonomy and preliminary investigation of social motivations. West. J. Speech Commun. **48**(4), 309–325 (1984)
10. Castelfranchi, C.: Artificial liars: why computers will (necessarily) deceive us and each other. Ethics Inf. Technol. **2**(2), 113–119 (2000)
11. Castelfranchi, C., Tan, Y.H.: Trust and Deception in Virtual Societies. Springer, Dordrecht (2001). https://doi.org/10.1007/978-94-017-3614-5
12. Castelfranchi, C., Tan, Y.H.: The role of trust and deception in virtual societies. Int. J. Electron. Commer. **6**(3), 55–70 (2002)
13. Chakraborti, T., Kambhampati, S.: (When) can AI bots lie? In: Proceedings of the 2019 AAAI/ACM Conference on AI, Ethics, and Society, pp. 53–59 (2019)
14. Chatila, R., Havens, J.C.: The IEEE global initiative on ethics of autonomous and intelligent systems. In: Aldinhas Ferreira, M., Silva Sequeira, J., Singh Virk, G., Tokhi, M., Kadar, E. (eds.) Robotics and Well-Being, pp. 11–16. Springer, Cham (2019). https://doi.org/10.1007/978-3-030-12524-0_2
15. Clark, M.H.: Cognitive illusions and the lying machine: a blueprint for sophistic mendacity. Ph.D. thesis, Rensselaer Polytechnic Institute (2010)
16. Coeckelbergh, M.: Are emotional robots deceptive? IEEE Trans. Affect. Comput. **3**(4), 388–393 (2011)
17. Coeckelbergh, M.: How to describe and evaluate "deception" phenomena: recasting the metaphysics, ethics, and politics of ICTS in terms of magic and performance and taking a relational and narrative turn. Ethics Inf. Technol. **20**(2), 71–85 (2018)
18. Cohen, P.R., Levesque, H.J.: Speech acts and rationality. In: 23rd Annual Meeting of the Association for Computational Linguistics, pp. 49–60 (1985)
19. Cohen, P.R., Perrault, C.R.: Elements of a plan-based theory of speech acts. In: Readings in Artificial Intelligence, pp. 478–495. Elsevier (1981)
20. Conger, J.A.: The necessary art of persuasion. Harv. Bus. Rev. **76**, 84–97 (1998)
21. Cushman, F.: Crime and punishment: distinguishing the roles of causal and intentional analyses in moral judgment. Cognition **108**(2), 353–380 (2008)
22. Danaher, J.: Robot betrayal: a guide to the ethics of robotic deception. Ethics Inf. Technol. **22**(2), 117–128 (2020)

23. De Rosis, F., Carofiglio, V., Grassano, G., Castelfranchi, C.: Can computers deliberately deceive? A simulation tool and its application to Turing's imitation game. Comput. Intell. **19**(3), 235–263 (2003)
24. Dignum, V.: Responsible Artificial Intelligence: How to Develop and Use AI in a Responsible Way. Springer, Cham (2019). https://doi.org/10.1007/978-3-030-30371-6
25. Dragan, A., Holladay, R., Srinivasa, S.: Deceptive robot motion: synthesis, analysis and experiments. Auton. Robot. **39**(3), 331–345 (2015)
26. Dunbar, N.E., Gangi, K., Coveleski, S., Adams, A., Bernhold, Q., Giles, H.: When is it acceptable to lie? Interpersonal and intergroup perspectives on deception. Commun. Stud. **67**(2), 129–146 (2016)
27. Evans, O., et al.: Truthful AI: developing and governing AI that does not lie. arXiv preprint arXiv:2110.06674 (2021)
28. Falcone, R., Castelfranchi, C.: Social trust: a cognitive approach. In: Castelfranchi, C., Tan, Y.H. (eds.) Trust and Deception in Virtual Societies, pp. 55–90. Springer, Dordrecht (2001). https://doi.org/10.1007/978-94-017-3614-5_3
29. Falcone, R., Singh, M., Tan, Y.H.: Trust in Cyber-Societies: Integrating the Human and Artificial Perspectives, vol. 2246. Springer, Heidelberg (2001). https://doi.org/10.1007/3-540-45547-7
30. Floridi, L.: Philosophy and Computing: An Introduction. Psychology Press (1999)
31. Floridi, L.: Ethics in the infosphere. Philosophers' Mag. **16**, 18–19 (2001)
32. Fogg, B.J.: Persuasive computers: perspectives and research directions. In: Proceedings of the SIGCHI Conference on Human Factors in Computing Systems, pp. 225–232 (1998)
33. Franklin, M., Ashton, H., Gorman, R., Armstrong, S.: Missing mechanisms of manipulation in the EU AI act. In: The International FLAIRS Conference Proceedings, vol. 35 (2022)
34. Greco, G.M., Floridi, L.: The tragedy of the digital commons. Ethics Inf. Technol. **6**(2), 73–81 (2004)
35. Habermas, J.: The Theory of Communicative Action: Lifeworld and Systems, a Critique of Functionalist Reason, vol. 2. Wiley, Hoboken (2015)
36. Häggström, O.: Strategies for an unfriendly oracle AI with reset button. In: Artificial Intelligence Safety and Security, pp. 207–215. Chapman and Hall/CRC (2018)
37. Hamblin, C.L.: Mathematical models of dialogue 1. Theoria **37**(2), 130–155 (1971)
38. Han, L., Siau, K.: Impact of socioeconomic status on trust in artificial intelligence (2020). AMCIS 2020 TREOs. 90. https://aisel.aisnet.org/treos_amcis2020/90
39. Hardin, G.: The tragedy of the commons. Science **162**(3859), 1243–1248 (1968)
40. HLEG, in AI: Ethics guidelines for trustworthy AI. B-1049 Brussels (2019)
41. Isaac, A., Bridewell, W.: White Lies on Silver Tongues: Why Robots Need to Deceive (and How). Oxford University Press, Oxford (2017)
42. Ishowo-Oloko, F., Bonnefon, J.F., Soroye, Z., Crandall, J., Rahwan, I., Rahwan, T.: Behavioural evidence for a transparency-efficiency tradeoff in human-machine cooperation. Nat. Mach. Intell. 1–5 (2019)
43. Kampik, T., Nieves, J.C., Lindgren, H.: Coercion and deception in persuasive technologies. In: 20th International Trust Workshop (co-located with AAMAS/IJCAI/ECAI/ICML 2018), Stockholm, Sweden, 14 July 2018, pp. 38–49. CEUR-WS (2018)
44. Kant, I.: On a supposed right to lie from philanthropy (1797). Practical philosophy [trans: Gregor m] (1996)
45. Leslie, A.M., Knobe, J., Cohen, A.: Acting intentionally and the side-effect effect: theory of mind and moral judgment. Psychol. Sci. **17**(5), 421–427 (2006)

46. Levine, E.E., Schweitzer, M.E.: Prosocial lies: when deception breeds trust. Organ. Behav. Hum. Decis. Process. **126**, 88–106 (2015)
47. Levine, T.R.: Encyclopedia of Deception. Sage Publications, Thousand Oaks (2014)
48. Levine, T.R.: Duped: Truth-Default Theory and the Social Science of Lying and Deception. University Alabama Press, Tuscaloosa (2019)
49. Lewis, P.R., Marsh, S.: What is it like to trust a rock? A functionalist perspective on trust and trustworthiness in artificial intelligence. Cogn. Syst. Res. **72**, 33–49 (2021)
50. Lippard, P.V.: "Ask me no questions, i'll tell you no lies";: situational exigencies for interpersonal deception. West. J. Commun. (includes Commun. Rep.) **52**(1), 91–103 (1988)
51. Masters, P., Smith, W., Sonenberg, L., Kirley, M.: Characterising deception in AI: a survey. In: Sarkadi, S., Wright, B., Masters, P., McBurney, P. (eds.) Deceptive AI. CCIS, vol. 1296, pp. 3–16. Springer, Cham (2020). https://doi.org/10.1007/978-3-030-91779-1_1
52. Mauldin, M.L.: Chatterbots, tinymuds, and the turing test: entering the loebner prize competition. In: AAAI, vol. 94, pp. 16–21 (1994)
53. Mell, J., Lucas, G., Mozgai, S., Gratch, J.: The effects of experience on deception in human-agent negotiation. J. Artif. Intell. Res. **68**, 633–660 (2020)
54. Miller, M.D., Levine, T.R.: Persuasion. In: An Integrated Approach to Communication Theory and Research, pp. 261–276. Routledge (2019)
55. Natale, S., et al.: Deceitful Media: Artificial Intelligence and Social Life After the Turing Test. Oxford University Press, Oxford (2021)
56. Ostrom, E.: Governing the Commons: The Evolution of Institutions for Collective Action. Cambridge University Press, Cambridge (1990)
57. Panisson, A.R., Sarkadi, S., McBurney, P., Parsons, S., Bordini, R.H.: Lies, bullshit, and deception in agent-oriented programming languages. In: Proceedings of the 20th International TRUST Workshop @ IJCAI/AAMAS/ECAI/ICML, pp. 50–61. CEUR Workshop Proceedings, Stockholm, Sweden (2018)
58. Panisson, A.R., Sarkadi, S., McBurney, P., Parsons, S., Bordini, R.H.: On the formal semantics of theory of mind in agent communication. In: Lujak, M. (ed.) AT 2018. LNCS, vol. 11327, pp. 18–32. Springer, Cham (2018). https://doi.org/10.1007/978-3-030-17294-7_2
59. Rawls, J.: A Theory of Justice. Harvard University Press, Cambridge (2009)
60. Resnick, D.: The Ethics of Science. Rout, London (1998)
61. Sætra, H.S.: Social robot deception and the culture of trust. Paladyn J. Behav. Robot. **12**(1), 276–286 (2021)
62. Sarkadi, S.: Deception. Ph.D. thesis, King's College London (2021)
63. Sarkadi, S.: An arms race in theory-of-mind: Deception drives the emergence of higher-level theory-of-mind in agent societies. In: 4th IEEE International Conference on Autonomic Computing and Self-Organizing Systems ACSOS 2023. IEEE Computer Society (2023)
64. Sarkadi, S., McBurney, P., Parsons, S.: Deceptive storytelling in artificial dialogue games. In: Proceedings of the AAAI 2019 Spring Symposium Series on Story-Enabled Intelligence (2019)
65. Sarkadi, S., Mei, P., Awad, E.: Should my agent lie for me? A study on attitudes of US-based participants towards deceptive AI in selected future-of-work scenarios. In: Proceedings of the 22nd International Conference on Autonomous Agents and Multiagent Systems (AAMAS 2023). IFAAMAS (2023)

66. Sarkadi, S., Panisson, A.R., Bordini, R.H., McBurney, P., Parsons, S.: Towards an approach for modelling uncertain theory of mind in multi-agent systems. In: Lujak, M. (ed.) AT 2018. LNCS, vol. 11327, pp. 3–17. Springer, Cham (2018). https://doi.org/10.1007/978-3-030-17294-7_1
67. Sarkadi, S., Panisson, A.R., Bordini, R.H., McBurney, P., Parsons, S., Chapman, M.D.: Modelling deception using theory of mind in multi-agent systems. AI Commun. **32**(4), 287–302 (2019)
68. Sarkadi, Ş, Rutherford, A., McBurney, P., Parsons, S., Rahwan, I.: The evolution of deception. R. Soc. Open Sci. **8**(9), 201032 (2021)
69. Sarkadi, S., Wright, B., Masters, P., McBurney, P. (eds.): DeceptiveAI, vol. 1296. Springer, Cham (2021). https://doi.org/10.1007/978-3-030-91779-1
70. Searle, J.R.: The Chinese room revisited. Behav. Brain Sci. **5**(2), 345–348 (1982)
71. Seiter, J.S., Bruschke, J., Bai, C.: The acceptability of deception as a function of perceivers' culture, deceiver's intention, and deceiver-deceived relationship. West. J. Commun. (includes Commun. Rep.) **66**(2), 158–180 (2002)
72. Sharkey, A., Sharkey, N.: We need to talk about deception in social robotics! Ethics Inf. Technol. **23**(3), 309–316 (2021)
73. Shim, J., Arkin, R.C.: A taxonomy of robot deception and its benefits in HRI. In: 2013 IEEE International Conference on Systems, Man, and Cybernetics, pp. 2328–2335. IEEE (2013)
74. Sklar, E., Parsons, S., Davies, M.: When is it okay to lie? a simple model of contradiction in agent-based dialogues. In: Rahwan, I., Moraïtis, P., Reed, C. (eds.) ArgMAS 2004. LNCS, vol. 3366, pp. 251–261. Springer, Heidelberg (2004). https://doi.org/10.1007/978-3-540-32261-0_17
75. Smith, W., Dignum, F., Sonenberg, L.: The construction of impossibility: a logic-based analysis of conjuring tricks. Front. Psychol. **7**, 748 (2016)
76. Sorensen, R.: Kant tell an a priori lie. In: From Lying to Perjury: Linguistic and Legal Perspectives on Lies and Other Falsehoods, vol. 3, p. 65 (2022)
77. Sorensen, R.A.: A Cabinet of Philosophical Curiosities: A Collection of Puzzles, Oddities, Riddles and Dilemmas. Oxford University Press, Oxford (2016)
78. Staines, P.: Linguistics and the Parts of the Mind: Or how to Build a Machine Worth Talking to. Cambridge Scholars Publishing, Cambridge (2018)
79. Turing, A.: Computing machinery and intelligence. Mind **59**(236), 433–460 (1950). www.jstor.org/stable/2251299
80. Van Maris, A., Zook, N., Caleb-Solly, P., Studley, M., Winfield, A., Dogramadzi, S.: Designing ethical social robots-a longitudinal field study with older adults. Front. Robot. AI **7**, 1 (2020)
81. Wagner, A.R., Arkin, R.C.: Robot deception: recognizing when a robot should deceive. In: 2009 IEEE International Symposium on Computational Intelligence in Robotics and Automation-(CIRA), pp. 46–54. IEEE (2009)
82. Wagner, A.R., Arkin, R.C.: Acting deceptively: providing robots with the capacity for deception. Int. J. Soc. Robot. **3**(1), 5–26 (2011)
83. Wang, D., Maes, P., Ren, X., Shneiderman, B., Shi, Y., Wang, Q.: Designing AI to work with or for people? In: Extended Abstracts of the 2021 CHI Conference on Human Factors in Computing Systems, pp. 1–5 (2021)
84. Weitz, K., Schiller, D., Schlagowski, R., Huber, T., André, E.: "do you trust me?" increasing user-trust by integrating virtual agents in explainable AI interaction design. In: Proceedings of the 19th ACM International Conference on Intelligent Virtual Agents, pp. 7–9 (2019)

85. Westlund, J.K., Breazeal, C.: Deception, secrets, children, and robots: what's acceptable. In: Workshop on The Emerging Policy and Ethics of Human-Robot Interaction, held in conjunction with the 10th ACM/IEEE International Conference on Human-Robot Interaction (2015)
86. Yudkowsky, E.: The AI-box experiment. Singularity Institute (2002)
87. Zhan, X., Xu, Y., Sarkadi, S.: Deceptive AI ecosystems: the case of chatgpt. In: Conversational User Interfaces, CUI 2023, 19–21 July 2023, Eindhoven, Netherlands (2023)

Neuro-Symbolic AI + Agent Systems: A First Reflection on Trends, Opportunities and Challenges

Vaishak Belle[1]([✉]), Michael Fisher[2], Alessandra Russo[3], Ekaterina Komendantskaya[4], and Alistair Nottle[5]

[1] University of Edinburgh, Edinburgh, UK
vbelle@ed.ac.uk
[2] University of Manchester, Manchester, UK
[3] Imperial College London, London, UK
[4] University of Southampton, Southampton, UK
[5] Airbus Central R&T AI Research Team, Bristol, UK

Abstract. To get one step closer to "human-like" intelligence, we need systems capable of seamlessly combining the neural learning power of symbolic feature extraction from raw data with sophisticated symbolic inference mechanisms for reasoning about "high-level" concepts. It is important to also incorporate existing prior knowledge about a given problem domain, especially since modern machine learning frameworks are typically data-hungry. Recently the field of neuro-symbolic AI has emerged as a promising paradigm for precisely such an integration.

However, coming up with a single, clear, concise definition of this area is not an easy task. There are plenty of variations on this topic, and there is no "one true way" that the community can coalesce around. Recently, a workshop was organized at AAMAS-2023 (London, UK) to discuss how this definition should be broadened to also consider reasoning about agents. This article is a collection of ideas, opinions, and positions from computer scientists who were invited for a panel discussion at the workshop.

This collection is not meant to be comprehensive but is rather intended to stimulate further conversation on the field of "Neuro-Symbolic Multi-Agent Systems."

1 Introduction

Artificial Intelligence (AI) is widely acknowledged as a new kind of science that will bring about (and is already enabling) the next technological revolution. Virtually every week, exciting reports come our way about the use of AI for drug discovery, game playing, stock trading and law enforcement. And virtually all of these are mostly concerned with a very narrow technological capability, that of predicting future instances based on past instances.

Fundamentally, the issue is that these models presuppose learning purely from data and treat features of interest often as independent and identically distributed random

The authors would like to express our gratitude to the organizers of the Neuro-symbolic Interest Group at the Alan Turing Institute for their encouragement and support.

variables. It is now widely acknowledged that this type of statistical information is limited in its ability to understand the world and model its knowledge [81]. Although there is no clear answer to how we might build systems that are more reliable, trust-worthy, and human-like, it has been suggested that this might involve a combination of symbolic constructs and data-driven machine learning [52], especially deep learning. Indeed, to get one step closer to "human-like" intelligence, we need systems capable of seamlessly combining the neural learning power of symbolic feature extraction from raw data with sophisticated symbolic inference mechanisms for reasoning about and learning "high-level" concepts [20,76]. It is important to also incorporate existing prior knowledge about a given problem domain [87], especially since modern machine learn-ing frameworks are typically data-hungry. Moreover, the development of systems that can simultaneously perform both neural and symbolic computations is crucial for the safe use of AI, as seen in the increasing interest in the verification of learning systems [101].

Despite this overarching motivation, coming up with a single, clear, concise defini-tion of Neuro-Symbolic AI is not an easy task. There are plenty of variations on this topic, and there is no "one true way" that the community can coalesce around.

Recently, a workshop was organized at AAMAS-2023 (London, UK) to discuss how this definition should be broadened to also consider reasoning about agents [46]. While much of machine learning usually focuses on single decisions that apply to recommen-dation and classification systems [89], it is reasonable to think that neuro-symbolic AI is targeting a more cognitive capacity, making the inclusion of agents a natural part of the equation. In fact, the common sense reasoning literature [36,70] typically involves an agent engaging in an environment where other users may also be interacting with the agent; therefore, considering multiple agents is also entirely natural.

The following is a collection of ideas, opinions, and positions from computer sci-entists who were invited for a panel discussion at the workshop, as well as feedback and interactions with the workshop's participants, which included both academics and industry experts. We do not claim that this collection is comprehensive, nor that we have discussed all the open issues and challenges in this area. The purpose of this article is to only present key insights that the panel felt needed to be expressed, with the hope that it will initiate a discussion. More generally, we hope this article will stimulate further conversation on the field of *Neuro-Symbolic Multi-Agent Systems*.

2 Key Considerations

Just as the field of neuro-symbolic AI is diverse – we invite readers to explore compre-hensive collections like [60] – the field of multi-agent systems is also broad and varied. Bringing together these two fields therefore brings even more variety and an expecta-tion that views across the disciplines will not necessarily be commonly held. At the outset, there are a few key considerations that are worth acknowledging. After all, sci-ence progresses once notions become precise. Given that the field is rapidly evolving, it is difficult to determine how these considerations should be addressed.

- Should we care that Neuro-Symbolic AI covers many varieties and is a "Broad Church"?

- Does it matter that there are different ways to achieve a capability? Is the current goal to build systems that simply perform a task, or do we need uniformity in how we approach and construct these systems? Is there a need for a unified mathematical language to bridge the gap between different system-building paradigms? Is diversity in paradigms hurting or helping the field?
- How can a balance between reasoning and learning be achieved in the context of agents and robots?
  - A long-standing question in the field of knowledge representation and cognitive robotics [74] is what kind of balance needs to be struck between the knowledge provided by the expert and what can be obtained from data. Data in itself may not immediately inform the underlying structure, so we need to consider how high-level concepts might be provided by humans that are then mapped onto low-level features. Can these high-level concepts themselves be perhaps partially or fully learned?
  - In practice, how much logic is actually needed? How much learning is actually needed? What kind of logic would make most sense? Classically, in neurosymbolic AI, we see the use of propositional logic [101] or finite-domain relational logic [125]. However, there is a motivation to consider temporal [65] and dynamic logic [97], particularly when contemplating agents operating within a physical environment.
  - Are there application areas where more "neuro" is needed and others where more "symbolic" is needed? Identifying the relationship between domains and applications, as well as the expectations in terms of expert knowledge vs structure learning from observational data could provide valuable guidance on how researchers should approach systems building.
- What can agent researchers learn from Neuro-Symbolic AI?
  - Agent systems researchers, as well as the multi-agent systems community, have long felt the need to integrate with machine learning to obtain observations about the real world [3] and other agents from data [127]. Clearly, neuro-symbolic offers an appropriate paradigm to think about these issues, although the notion of combining high-level control and low-level perception has been an established line of inquiry over the last few decades [97].
- Which industry is ready for Neuro-Symbolic AI?
  - As mentioned above, one-shot learning for predictions and classifications does not seem to require cognitive capabilities such as reasoning [89], although they can still be useful for tasks like visual question-answering and respecting domain constraints [107]. On the other hand, agents systems, where a model of the world needs to be explicitly maintained, present an obvious case for integrating symbolic structures and reasoning. Suggesting a set of application domains would aid in this area.
- Is Large Language Models the answer?
  - Large language models have demonstrated a surprising range of capabilities [71], although their reliability and consistency are still lacking [8]. Nonetheless, they seem capable of harnessing textual data to reason about common sense, a longstanding issue in this field [98]. Therefore, what specific ways can these

models encompass some of the abilities outlined by (neuro) symbolic AI? Alternatively, how might combining symbolic AI with deep learning become significant in obtaining consistent and reliable answers [29,49]?

– What are the barriers to using Neuro-Symbolic AI? What issues need to be addressed next?

  • A long-standing barrier to symbolic AI has been the lack of experts and the translation of expert knowledge into consistent theories of logical knowledge. Might similar issues also be a barrier for Neuro-symbolic AI? It should be, however, noted that there are approaches for learning symbolic knowledge from data [73]. What other concerns might we have? For example, scalability and the lack of differentiability in classical logic might be other issues, although steps are being taken to tackle these [45,51,61].

In what follows, we provide excerpts from the panel discussion regarding the responses to some of the above questions. The answers are not meant to be exhaustive. We merely offer a glimpse into some of the thoughts that have emerged regarding these questions. It should be noted that certain responses are longer than others, potentially indicating areas that have seen more development. We have abbreviated the names of the authors using initials.

3   Should We Care that Neuro-Symbolic AI Is a "Broad Church" Covering Many varieties?

As the name suggests, Neuro-Symbolic AI brings together two different fields of AI: the more connectionist approaches (the "Neuro") and the more high-level, human-relatable perspective (the "symbolic"). These contrasting fields already imply that "Neuro-Symbolic AI" incorporates input from a wide range of backgrounds. The central concern here is whether this diversity could pose difficulties in the maturation and development of Neuro-Symbolic approaches in a coherent and productive manner.

However, it is important to acknowledge that there is also strength in this diversity. Different approaches allow for the sharing of best practices and more creative problem-solving methods, eliminating the need to constantly "reinvent the wheel". As is usual in computer science, areas typically start off with a lot more diversity and ad-hocness until some ideas mature. But ultimately, a careful balance between diversity and coherence must be achieved for systems to be useful and adopted.

MF: It is important to incorporate all varieties as each has its place [52]. For example, tightly coupled neuro-symbolic models (with combined semantics) provide important capabilities around explainability and transparency [51], while loosely-coupled neuro-symbolic approaches are particularly useful in practical systems architectures [78,112,125]. And, while the former suffers from the need to have probabilistic models (and so weak verification) the latter is not able to provide a holistic/comprehensive semantics for all components. Consequently, while neither extreme is foolproof, many combinations have important uses.

AR: We now need more than ever a wide range of AI solutions capable of addressing the global challenges that our society is facing [39]. These challenges come with their own complexity and characteristics, for which targeted solutions are likely to be more effective than a single general method that fits all. We require a broad range of neuro-symbolic AI solutions capable of performing various human-like cognitive tasks to appropriately support humans [30]. These tasks include prediction, planning, forecasting, explanation generation, counterfactual inference, argument construction, generalization, and generation of new hypotheses, among others.

Each of these cognitive tasks has been well studied in the field of symbolic AI, with their formalization, computational characteristics, and complexity thoroughly explored. Numerous targeted solutions and methods have been developed. Symbolic AI has a lot to offer to the purely statistical AI community, which has been primarily focused on data rather than cognition, symbolic grounding, generalization, and abstraction [20].

The integration of neuro and symbolic AI will achieve its best outcomes if allowed to grow through a broad range of expertise, knowledge, technologies, and best practices.

VB: I don't think we should care about the lack of cohesiveness at the moment. As mentioned above, our society is facing urgent problems, and seeking novel and diverse solutions is important.

Nonetheless, in terms of semantics [112] and correctness [72], striving for a mathematical framework that can concretely and formally unify different approaches is valuable. We can consider advancements in the theory of databases [77] and the semantic web [83] in terms of how the mathematical foundations for these fields have facilitated healthy growth.

It is also important to contemplate the potential of logic to serve as a meta-theory [44] that provides a formalization framework for integrating various machine learning components and facilitating user interaction [15,54]. Epistemic logic, for example, have been extensively employed in studying cryptography [59], distributed systems [58], and game theory [7], as they allow for the representation of multiple agents and their mental states. Such constructions could prove useful when dealing with explanations and fairness in increasingly complex systems [13,119]. This idea has been particularly prominent in explainable AI planning [102].

4   How Can a Balance Between Reasoning And learning Be Achieved in the Context of Agents / robots?

MF: To a large extent, the balance between the neuro- and symbolic aspects depends on the uses, particularly the risks involved in the technologies. If there are very limited requirements for strong verification or full explanations, then a majority of neural and data-driven approaches may be appropriate. Alternatively, if a safety-critical system is involved, and verification [38,101] and explanation [103] become central, then symbolic approaches are more clearly supported.

AR: Reasoning, if taken in its general sense (e.g. the ability to reach conclusions from premises and facts) [110], should be key for any agent/robot, whether it be human or machine [20]. An agent with this skill has the ability to adapt its behavior more easily to different situations where different facts are present and different conclusions need to be reached. Obviously, the level of complexity of the task at hand determines the level of complexity of the reasoning needed. Agents based on pure data-driven learning methods (e.g. RL) will always suffer from the inability to reuse their skills and learned knowledge to solve new tasks, as they are not capable of "logically" constructing solutions from what they know and what they observe.

VB: The balance between logic and learning can be achieved by combining both approaches in a way that leverages the strengths of each. For example, logic can be used to model complex systems and interpret data, as observed in statistical relational AI [93]. Additionally, learning can be used to adapt to new situations and improve performance over time [47].

In the context of agent systems, logic is commonly employed for ontologies and temporal properties, which could play a role for knowledge-enhanced learning agents [27,28] and dynamic modelling [19,65]. It enables us to model agents with beliefs and intentions [57], which may not be easily accomplished with neural networks alone.

Recent advancements in Neuro-symbolic AI frameworks have demonstrated that even the incorporation of small logical constraints can significantly enhance the performance of semi-supervised learning [51,61]. However, it is important to note that these neural networks do not always guarantee exact satisfaction of the logical constraints.

Generally, the lack of guarantees in neural networks can be significant for certain applications, particularly those where safety and reliability are critical factors to consider. With large language models, we might also have wildly inaccurate suggestions [8]. In such cases, relying solely on neural networks may not be sufficient, and incorporating logical reasoning can provide more robustness.

5   In Practice: How Much Logic Is Really needed? How Much Learning Is Really needed?

MF: As for the previous question, whenever it is necessary to ensure robustness in agent decision-making [38], emphasis will be placed on symbolic approaches [24]. However, in situations where dealing with unexpected and unstructured data is more important, neural techniques will be prioritized.

AR: It is important to note that learning is not necessarily just a data-driven neural computation process. For instance, symbolic learning is a key example of symbolic AI that aims at learning abstract knowledge, hypotheses, and definitions from observations [93] and examples [87]. So a better question would be how much logic or neural computation is really needed? AI systems should combine all three elements of data-driven learning (or better pattern matching and feature extraction), reasoning, and symbolic learning. The former is because information often comes in multiple forms (sensing, imaging, textual) and less in symbolic structural form [82]. Symbolic learning is crucial because the knowledge needed to perform symbolic reasoning might not be known a priori and might need to be learned (as general knowledge) from the data-driven learned outcomes and observations [109]. Consider, for instance, the case of deciding how to treat patients who suffer from a new pathogen (e.g., long COVID). Existing clinical knowledge about the specifics of the new pathogen is very limited and subject to change, based on new clinical evidence and effects of decision outcomes. But it can be learned to "augment" (being symbolic) clinical knowledge and help in performing improved informed decisions [27]. Therefore, symbolic learning should always be a key component in an AI system to abstract data-driven learned outcomes into general knowledge. The amount of data-driven learning and symbolic reasoning depends on the task at hand.

VB: The short answer is that this may very well depend on the situation.

For a longer answer, it is worth reflecting on the nature of knowledge representation and how it affects modeling and learning. Regardless of whether we use a logical representation (such as a logic program), a probabilistic one (such as a Bayesian network), or a connectionist one (such as a neural network), our goal is to represent information and capture the knowledge of a robot. We want to understand how this representation produces interesting conclusions that determine what the robot knows and understands about the world [20].

While acquisition is undoubtedly important, we need to also consider the language used for representing information. In a statement remarkably similar in spirit, Pearl writes [91]:

This is why you will find me emphasizing and reemphasizing notation, language, vocabulary and grammar. For example, I obsess over whether we can express a certain claim in a given language and whether one claim follows from others. My emphasis on language also comes from a deep conviction that language shapes our thoughts. You cannot answer a question that you cannot ask, and cannot ask a question that you have no words for.

Moreover, from the perspective of explainability and trustworthiness, we require a language that is understandable to domain experts and allows for the updating and provision of new information, while also being suitable for computational reasoning.

We might further draw a distinction between *explicit* and *implicit* knowledge: explicit knowledge is often the information that is directly modeled or provided or learned. This could come in the form of rules, databases, knowledge bases, graphs, or any other structured data. Implicit knowledge is what is obtained from the explicit knowledge through one or more reasoning steps.

The question then becomes: how much implicit knowledge is really required for the application at hand, and how reliably can a machine learning system produce it? For example, large language models seem to generate plausible logical conclusions in certain situations, but not consistently [29,49]. Therefore, at this stage, we need a logical solver of some kind that is capable of performing logical reasoning. Lastly, it is worth reflecting on the key variations of logical reasoning, which may need to be enabled in a principled manner in Neuro-symbolic AI in the future:

– *Deduction:* from sentences $\alpha, \beta, \ldots, \gamma$, if all permutations of truth assignments to these sentences also make true a sentence δ, then we write $\alpha \wedge \beta \wedge \ldots \gamma \models \delta$. For example, if we know humans are mortal, and Bob is a human, then we conclude Bob is mortal too.

– *Induction:* from sentences $\alpha(1), \alpha(2), \ldots, \alpha(n)$, if we can identify a sub-formula (or pattern) $\beta(x)$, such that it describes every observation we make of the world, then we say $\beta(x)$ is an induced formula that captures the instances in $\alpha(x)$. For example, if we see that the parent of the parent of Bob is referred to as the grandparent of Bob, and likewise for Lisa, and so too for John, we can conclude that $parent(x, y) \wedge parent(y, z) \rightarrow grandparent(x, z)$.

– *Abduction:* given sentence α and observation β, what must be the smallest assumption δ such that $\alpha \wedge \delta$ explain β? This, of course, is the Sherlock Holmes strategy for solving cases, mistakenly referred to as deduction by Arthur Conan Doyle.

6 What Can Agent Researchers Learn From Neuro-Symbolic AI?

MF: Particularly in the area of multi-agent systems, the loose coupling of neural and symbolic components has a close analogy with heterogeneous multi-agent systems [122]. Just as in hybrid agent architectures for autonomous robotics [75], we can utilize a range of different types of agents in our multi-agent systems, often with differing balances of neural and symbolic components. These heterogeneous systems, while more difficult to formalize and analyze, are often more flexible and appropriate for complex and uncertain environments.

AR: Fundamentally, the fact that an intelligent agent should be seen as a system of multiple components (e.g., perception, cognition, and action) working together to solve a task is important to consider [86]. The integration of these components, or skills, is exactly what neuro-symbolic AI essentially offers. In the context of multi-agent systems, the cognitive ability of the agents can aid communication and foster improved collaboration, particularly in the context of human-agent collaboration [30,67,118].

7 Are There application Areas Where We Clearly Need More "neuro" and Are There others Where We Clearly Need More "symbolic"?

MF: As mentioned above, when strong verification or clear and justified explanations are needed, we would anticipate a focus on the "symbolic" aspect. This is primarily due to the fact that both verification and explainability [5] heavily rely on transparency, and high levels of transparency are commonly found in symbolic components [96], but not typically in neural components.

AN: When it comes to explainability there are also considerations of *interpretability* as well as *transparency* and *intelligibility*. All these factors are strongly influenced by symbolic approaches, and are thus very important for achieving verification and certification. Potentially opaque neural approaches are not well suited.

Explainable AI (XAI) has been focussed on more *post-hoc* techniques and methodologies - i.e. taking high-performing 'black box' systems which offer little in the way of transparency or interpretability and providing specific explanations. This approach calls for an understanding of 'neuro' in the original approach, and a strong application of 'symbolic' in the explanatory approach. A greater reliance towards more cognitively inspired approaches, would allow for a more *inherently explainable* form of AI and help drive development towards *'the ability to explain or to present in understandable terms to a human'* [40].

8   Which Industry Is "ripe" For Neuro-Symbolic AI?

MF: My own experience is in practical autonomous systems [22], and particularly with the increasing need for strong verification or clear and justified explanations [23], this seems to be a prime candidate for an area where neuro-symbolic approaches can have an impact. Improving verification evidence to a regulator, along with enhancing explanations to a user, would both seem to be important.

AR: I believe that healthcare and clinical decision-making are particularly suited for novel solutions in neuro-symbolic AI [96]. Features such as explainability, interpretability, the ability to generalize to out-of-distribution data, and the ability to adapt learned knowledge (through symbolic learning) are key to ensuring an effective and positive adoption of AI [37], with the potential to reduce human errors. Neuro-symbolic AI holds the potential to bring about the transformative change that the field of digital healthcare has been waiting for, for a long time.

AN: Within the aerospace industry there is much discussion on how to introduce data-driven approaches into products in a way that only enhances safety. The European Union Aviation Safety Agency (EASA) see explainability as one of the three main components of trustworthy AI [42], and place human-centricity within their roadmap [41].

9   Is the Answer Large Language Models?

MF: Depends what the question is! The fact that neither strong verification nor clear explanations can be provided for LLMs should restrict their use in critical systems.

AR: Not yet. Very little is still known about the "hows and whys" of these models in order to safely use them to tackle tasks that have social and economic impact [37]. They need to be further studied, and mechanisms for controlling, constraining, and verifying them need to be developed first to see them playing a key role in the broad use of AI in our society.

AN: The power, and recent popularity, of Large Language Models is undeniable! To assume that they are the answer to everything is a little premature, particularly as the processes necessary for robust verification and explanation are still in their infancy.

We researchers should also take guidance from some of the reasons that LLMs have proved so popular. For many, the attraction of a LLM isn't so much the underlying technical implementation, but rather the *user experience and interaction* of these AI systems. The ability to interact with complex systems in a natural language manner, and receive clear, unambiguous results are a key driver for their popularity and acceptance. There is much that the field of Neurosymbolic AI can bring towards making AI systems more naturally intuitive, making for greater acceptance and more 'useful' AI. Engineers are well used to the concept of requirement driven development, where it is important to first capture the user needs, and this needs to extend to AI systems, for instance in ensuring that AI explanations meet the needs of the specific user [56].

VB: A bit of context might be useful here. There are many areas of overlap between language, logic, and learning [43]. Naturally, then, with the recent prominence of chatbots powered by large language models, there is quite a lot of chatter in terms of how much these models impact the fundamental science of artificial intelligence. For instance, is it the case that they will lead to new models of language acquisition and linguistic reasoning? What kind of logical and mathematical capabilities will they have?

It is important to note that the basic architecture seems most equipped to pick up on statistical correlations by predicting the subsequent word based on a previous set of words. Usually, the models predict the next word by utilizing stochastic associations and a hint of randomness to lend creativity to the prediction. What makes them particularly noteworthy, compared to most language models from, say, even late as 2016, is that their particular attention mechanism. This is enabled via the so-called "transformer" architecture [116], which allows them to keep track of the context for, say, a 100 or even a 1000 words, thereby making the next predicted word a very good suggestion (from a probability viewpoint). This makes them so powerful that they can, say, summarise a paragraph such as the one I just wrote (or even a paper), rephrase it in the style of, say, Shakespeare.

However, a number of early inquiries are showing that these models are flawed in understanding relationships, symmetries, planning and reasoning [114]. Of course, it is impossible to predict how this technology will evolve in the coming years, but it seems reasonable to believe that much of the causal and symbolic knowledge of the world cannot be picked up purely by statistical associations [91]. Even if many of them eventually turn to be acquired from data, to ensure safety and critical applications, we would need a final layer of symbolic reasoning and verification. And such a model could very well be within the paradigm of neuro-symbolic AI.

For instance, recently, Wolfram Alpha has formed a formal partnership with OpenAI's chatGPT, through which it integrates WolframAlpha's symbolic reasoning over the outputs of the large language models for reliable mathematical reasoning [121]. This integration aligns closely with the principles of neuro-symbolic AI.

Ultimately, large language models do not possess an explicit world model, and it is not clear if one that can be explicitly extracted from its parameter space. This absence of a grounded understanding of the world makes their knowledge superficial and unreliable. Connecting linguistic constructs with structured knowledge, such as knowledge bases or ontologies, will provide the necessary element for creating an autonomous agent that has an account of the world that it operates in. A model of beliefs and intentions will ensure that its actions can be interpreted against explicit (presumably human-given) goals. This approach too aligns with the principles of neuro-symbolic AI.

10 What Are the Barriers to Using Neuro-Symbolic AI? What Issue Needs to Be Fixed next?

MF: Understanding the full breadth of neuro-symbolic approaches to agents and categorizing each of their strengths and drawbacks will surely be important. We are only at the beginning of this endeavor.

AR: There are many barriers to overcome. Key barriers include the need for finding better ways to integrate the continuous and discrete nature of neuro and symbolic AI [45,51,80]. The mathematical principles of this integration still need to be fully understood and solved in a way that overcomes their respective limitations (e.g., the vanishing gradient problem for neural components, scalable methods for symbolic reasoning and learning).

11   Current Trends in Neuro-Symbolic AI

Before concluding this article, we wish to briefly present a view of current exciting trends in neuro-symbolic AI as seen by some of the panel members. Of course, this is undoubtedly a reflection of the interest and biases of the panelists, as is the rest of this article.

VB: These are my personal favorites, and therefore, I do not intend to disrespect the many other directions not discussed here. (In fact, the interested reader should check out collections such as [52,60] to get a glimpse of all the breadth of the area).

- **Knowledge graphs.** Topics here range from learning such graphs through neural approaches, as well as reasoning about them, either using a logical approach or a neural one [17]. Knowledge graphs have been used to model protein databases [94], social networks [4], as well as commonsense knowledge in advanced language models [98].
- **Neuro-symbolic logic programs.** This can be seen to extend knowledge graphs with non-trivial axioms as specified in a logic program. DeepProbLog [80], for example, interfaces neural predicates as external artifacts in probabilistic logic programs. Logic Tensor Networks [10] permit neural outputs to be artifacts in first-order theories. There is also work on using a compositional semantics for such programs [112].
- **Differential program induction.** Program induction is a classic problem in computer science [55]. Using neural techniques allows for a fresh perspective on the problem. Approaches have been investigated to learn concepts in programs [73], rules in a relational logic program [45], as well as the policies of a reinforcement learning agent [14].
- **Training neural networks with logic formulas.** This sub-area is about constraining the loss function of neural networks with logical formulas so as to ensure that the predictions of neural network respect domain constraints. There have been a number of variants over the years [51,61,107], all of which are also motivated by the problem of label-free supervised learning. A special twist of this problem in a dynamic setting is the idea of rewarding reinforcement learning agents with temporarily extended goals, as seen in the work of reward machines [65].

KK: In AI literature, the term "neuro-symbolic reasoning" refers broadly to any combination of logical and statistical reasoning. There are several survey papers that make an attempt to categorise different trends in the field [16,35,53]. Examples of neuro-symbolic reasoning include incorporating symbolic knowledge into neural network training [9], embedding neural networks into logical inference [120], learning (probabilistic) logic programs or datalog programs [11,31,106]. Here are a few topics of interest for me in particular.

Differentiable Logics: As an example, a specific branch of neuro-symbolic reasoning focuses on the generation of loss functions (for neural networks or other learning systems) based on logical specifications. These translations are also referred to as "differentiable logics" or DLs. The initial attempt to translate propositional logic specifications into loss functions was presented in [124], and later extended to a subset of first-order logic in [48]. Furthermore, this work was enhanced by introducing a fuzzy interpretation to DL in [115]. [105] proposed a generalization for the syntax and semantics of DLs, aiming to incorporate all previously introduced DLs into one formal system and compare their theoretical properties.

Neuro-symbolic Programming: More recently, neuro-symbolic programming was introduced as a separate term by [26] and defines a set of principles for writing programs that "combine elements of both neural networks and classical symbolic programs". Initially, the term was introduced in [25], as a "generalisation of classic program synthesis that includes machine learning components".

Neural Network Verification: One of the first known attempts to verify neural networks was based on abstract interpretation [92]. The famous paper by [108] highlighted the problem of neural network vulnerability to small input perturbations ("adversarial attacks") and gave additional impetus to this line of research. In CAV-2017, two papers, by [63,68], on neural network verification appeared, and both utilized specialized forms of SMT-solving.

The latter gave rise to Marabou [69], a rapidly developing sound and complete neural network verifier. In 2019, the ERAN verifier by [104] appeared in POPL and showed that the performance of abstract interpretation methods compares favorably against those based on SMT solving. However, the range of properties they handled was limited to proving the adversarial robustness of neural networks. Moreover, ERAN was incomplete. This line of research led to many subsequent extensions, such as those by [88,90], to mention a few. Mixed Integer Linear Programming (MILP) methods were introduced into this community by [12] and were further developed into working tools, such as Venus by [18]. Neural network verifier extensions followed two main directions: scaling to larger networks – $\alpha\beta$-Crown [117] and GCP-Crown [128], the winners of VNN-COMP (verifying neural networks competition) [21] in 2021 and 2022 are worth mentioning – and extending from piece-wise linear to non-linear activation functions. An example of that is the handling of sigmoid neurons by Verisig using interval arithmetic [66].

At the time of writing, there exist over a hundred verifiers for neural networks. Several papers and monographs are dedicated to the survey of the landscape [2,62,79]. The community established specification standards (https://www.vnnlib.org/[1]), common benchmarks, and annual competitions.

AR: My view, as expressed above, is that to get one step closer to "human-like" intelligence, we need systems capable of seamlessly combining the neural learning power

---
[1] <https://www.vnnlib.org/>.

of symbolic feature extraction from raw data with sophisticated symbolic inference mechanisms for reasoning about and learning "high-level" concepts. This should also incorporate existing prior knowledge about a given problem domain. The development of systems that can perform both neural and symbolic computations simultaneously is crucial for the safe, trusted, and responsible use of AI. Here are a few developments that I believe are exciting to look into further.

Neuro-symbolic Reasoning: Quite a few examples of neural-symbolic architectures have been proposed. Novel model architectures have been introduced to learn the model of a program (e.g. [85]), symbolic rules [45], and the reasoning process itself [100]. These approaches perform approximate inference and typically depend on replacing exact logic with fuzzy counterparts. The enhancement of neural network training with symbolic constraints has also been explored [123].

However, this treatment considers high-level reasoning only as a guide to the neural training process, not as a goal in itself. In recent years, approaches that integrate low-level neural processing with high-level symbolic inference in an end-to-end fashion have been developed [99]. These approaches are seen as a method for enhancing a neural network by providing it with some of the reasoning capabilities and interpretability that a symbolic system traditionally possesses.

Many of these systems are built around differentiable neural architectures that learn to approximate symbolic computation from data [84, 95]. Others extend the capabilities of symbolic solvers by allowing the system to depend on the output of a neural component, and using inconsistent results from the neural output, reasoning process and data to guide the training of the network. They see neuro-symbolic reasoning on raw data as a combination of two tasks: a perception task, handled by a neural component, and a reasoning downstream task, handled by the symbolic solver [1, 126]. Their advantages include the ability to train a neural network by only using labels of a given downstream task, increased interpretability, and the ability to generalize to downstream tasks other than the one the system is trained on.

However, their training time can be significantly long, even when the perception task is simple [113, 126]. This is because, in order to train the neural component, the symbolic component has to "work backwards" from a given downstream label to a set of possible latent labels, which can be exponential in size. This fact has, so far, hindered most neuro-symbolic reasoning systems from being applied to tasks involving a larger number of inputs. Additionally, more complex perception tasks are usually not considered, with a few exceptions, e.g. [64].

Scalable neuro-symbolic reasoning approaches capable of producing neuro-symbolic models that are interpretable and generalize beyond the training task have recently been proposed. An example is the work in [6], which trains a neural perception module and a reasoning component on the downstream task end-to-end. This approach avoids the combinatorial difficulty faced by other neuro-symbolic reasoning systems. Furthermore, it analyzes the clustered structure of the learned embedding space by identifying latent concepts and assigning them symbolic meaning through an optimization task. This task is solved using a symbolic system and background knowledge of the task. This transformation lifts the purely neural model into a neuro-symbolic one that is interpretable and capable of generalizing to tasks beyond the capabilities of the neural network alone.

Neuro-symbolic Learning: One of the main drawback for neuro-symbolic reasoning approaches is that they require a complete and manually engineered representation of the symbolic knowledge related to the task in hand. On the other hand, neuro-symbolic learning systems combine neural components and symbolic learning to significantly reduces the amount of engineering and labelling required. [33] extends [34] by using abduction and induction to jointly train a neural network and learn symbolic knowledge from raw data. But, it suffers from the fact that the network is vulnerable to becoming stuck in a local optima when the space of latent concepts for the neural prediction is very dense with respect to the downstream label. [32] overcomes this problem. It trains a general neural network to classify latent concepts from raw data, whilst learning an expressive and interpretable knowledge to solve computationally *complex* problems. It uses a symbolic learner to learn a logic program and trains the neural network by *reasoning* over the learned knowledge and using a semantic loss function.

Neuro-symbolic Reinforcement Learning: Neuro-symbolic reasoning has also been proposed in the context of agents and reinforcement learning, with the objective of addressing three main shortcoming of pure reinforcement learning methods: (1) inability to generalise outside the task they are trained, (2) lack of interpretability and (3) data inefficiency. [86] proposes a *DUA* architecture composed of a *Detect* component that extracts symbolic object representations from the raw data of the environment using computer vision, an *Understand* component that learns a symbolic meta-policy over options using symbolic learning, and an *Act* component that uses individually trained options. The architecture may be loosely thought of as a two-systems solution: the options represent the fast, reactive and non-interpretable facets of an intelligence agent, while the symbolic meta-policy learning is the substrate of the slow, logically rational and interpretable side of the agent's intelligence.

Abstraction is key to achieve generalisation and transfer in RL agents. Dividing a single task into several subtasks that can be solved separately is among the promising approaches to abstracting RL. Abstract hierarchies of tasks can be expressed symbolically in the form of automata. They can be engineered [111], similarly to how symbolic knowledge is engineered in neurosymbolic reasoning methods described above, or they can be learned [50] from observation traces perceived by an RL agent. The learned automata capture the agent's task decomposition into a structure of subtaks. The agent's RL algorithm can at the same time exploit this task structured decompostion to learn the policies needed to accomplish the subtasks. Multiple subpolicies can therefore be learned simultaneously and reused in multiple tasks thus realising knowledge transfer across multiple tasks. The RL and automaton learning processes are interleaved: the agent can immediately leverage a new (partially correct) learned subgoal automaton and when a trace is not correctly recognized by the automaton, a new automaton is symbolically learned.

12   Conclusions

In this article, we essentially aimed to provide an overview of the challenges and opportunities of neuro-symbolic AI from agent-systems research perspective. We believe that many of these challenges are of course inherited from symbolic learning and statistical relational learning. However, the use of deep learning and large neural networks introduces a higher level of complexity. This complexity arises from the need to scale with large high-dimensional data, while also addressing issues such as lack of robustness, a vast hyper-parameter space, and limited explainability due to their black box nature. We hope that these discussions will be valuable to researchers with an interest in this field, and that this article will initiate a dialogue on the direction of future research.

References

1. Manhaeve, R., Dumančić, S., Kimmig, A., Demeester, T., De Raedt, L.: Neural probabilistic logic programming in deepproblog. Artif. Intell. **298**, 103504 (2021)
2. Albarghouthi, A.: Introduction to neural network verification. Found. Trends ® Program. Lang. **7**(1–2), 1–157 (2021)
3. Albrecht, S.V., Stone, P.: Autonomous agents modelling other agents: a comprehensive survey and open problems. Artif. Intell. **258**, 66–95 (2018)
4. Anderson, C., Domingos, P., Weld, D.: Relational Markov models and their application to adaptive web navigation. In: Proceedings of SIGKDD, pp. 143–152. ACM (2002)
5. Arrieta, A.B., et al.: Explainable artificial intelligence (XAI): concepts, taxonomies, opportunities and challenges toward responsible AI. Inf. Fusion **58**, 82–115 (2020)
6. Aspis, Y., Broda, K., Lobo, J., Russo, A.: Embed2sym-scalable neuro-symbolic reasoning via clustered embeddings. In: Proceedings of the International Conference on Principles of Knowledge Representation and Reasoning, vol. 19, pp. 421–431 (2022)
7. Aumann, R.J.: Interactive epistemology II: probability. Int. J. Game Theory **28**(3), 301–314 (1999)
8. Azamfirei, R., Kudchadkar, S.R., Fackler, J.: Large language models and the perils of their hallucinations. Crit. Care **27**(1), 1–2 (2023)
9. Badreddine, S., d'Avila Garcez, A.S., Serafini, L., Spranger, M.: Logic tensor networks. Artif. Intell. **303**, 103649 (2022)
10. Badreddine, S., Garcez, A.D., Serafini, L., Spranger, M.: Logic tensor networks. Artif. Intell. **303**, 103649 (2022)
11. Bárány, V., ten Cate, B., Kimelfeld, B., Olteanu, D., Vagena, Z.: Declarative probabilistic programming with datalog. ACM Trans. Database Syst. **42**(4), 22:1-22:35 (2017)
12. Bastani, O., Ioannou, Y., Lampropoulos, L., Vytiniotis, D., Nori, A.V., Criminisi, A.: Measuring neural net robustness with constraints. In: Lee, D.D., Sugiyama, M., von Luxburg, U., Guyon, I., Garnett, R. (eds.) Advances in Neural Information Processing Systems 29: Annual Conference on Neural Information Processing Systems 2016, December 5-10, 2016, Barcelona, Spain, pp. 2613–2621 (2016)
13. Belle, V.: Knowledge representation and acquisition for ethical AI: challenges and opportunities. Ethics Inf. Technol. **25**(1), 22 (2023)
14. Belle, V., Bueff, A.: Deep inductive logic programming meets reinforcement learning. In: The 39th International Conference on Logic Programming. Open Publishing Association (2023)
15. Belle, V., De Raedt, L.: Semiring programming: a semantic framework for generalized sum product problems. Int. J. Approx. Reason. **126**, 181–201 (2020)

16. Besold, T.R., et al.: Neural-symbolic learning and reasoning: a survey and interpretation. In: Hitzler, P., Sarker, M.K. (eds.), Neuro-Symbolic Artificial Intelligence: The State of the Art, vol. 342. Frontiers in Artificial Intelligence and Applications, pp. 1–51. IOS Press (2021)

17. Bosselut, A., Le Bras, R., Choi, Y.: Dynamic neuro-symbolic knowledge graph construction for zero-shot commonsense question answering. In: Proceedings of the AAAI Conference on Artificial Intelligence, vol. 35, pp. 4923–4931 (2021)

18. Botoeva, E., Kouvaros, P., Kronqvist, J., Lomuscio, A., Misener, R.: Efficient verification of Relu-based neural networks via dependency analysis. In: The Thirty-Fourth AAAI Conference on Artificial Intelligence, AAAI 2020, The Thirty-Second Innovative Applications of Artificial Intelligence Conference, IAAI 2020, The Tenth AAAI Symposium on Educational Advances in Artificial Intelligence, EAAI 2020, New York, NY, USA, February 7-12, 2020, pp. 3291–3299. AAAI Press (2020)

19. Boutilier, C., Reiter, R., Price, B.: Symbolic dynamic programming for first-order MDPs. In: Proceedings of IJCAI, pp. 690–697 (2001)

20. Brachman, R.J., Levesque, H.J.: Machines Like Us: Toward AI with Common Sense. MIT Press, Cambridge (2022)

21. Brix, C., Müller, M.N., Bak, S., Johnson, T.T., Liu, C.: First three years of the international verification of neural networks competition (VNN-comp). Int. J. Softw. Tools Technol. Transf. **25**, 1–11 (2023)

22. Charisi, V., et al.: Towards moral autonomous systems. arXiv preprint arXiv:1703.04741 (2017)

23. Chatila, R., et al.: Trustworthy AI. Reflect. Artif. Intell. Human. **2021**, 13–39 (2021)

24. Chatterjee, K., Chmelik, M., Gupta, R., Kanodia, A.: Qualitative analysis of Pomdps with temporal logic specifications for robotics applications. In: 2015 IEEE International Conference on Robotics and Automation (ICRA), pp. 325–330. IEEE (2015)

25. Chaudhuri, S., Ellis, K., Polozov, O., Singh, R., Solar-Lezama, A., Yue, Y.: Neurosymbolic programming. Found. Trends Program. Lang. **7**(3), 158–243 (2021)

26. Chaudhuri, S., Solar-Lezama, A., Sehgal, A.: Neurosymbolic programming. Found. Trends Program. Lang. **7**, 148–153 (2023)

27. Confalonieri, R., Weyde, T., Besold, T.R., del Prado Martín, F.M.: Using ontologies to enhance human understandability of global post-hoc explanations of black-box models. Artif. Intell. **296**, 103471 (2021)

28. Cozman, F.G., Munhoz, H.N.: Some thoughts on knowledge-enhanced machine learning. Int. J. Approx. Reason. **136**, 308–321 (2021)

29. Creswell, A., Shanahan, M., Higgins, I.: Selection-inference: exploiting large language models for interpretable logical reasoning. arXiv preprint arXiv:2205.09712 (2022)

30. Crootof, R., Kaminski, M.E., Price II, W.N.: Humans in the loop. Vanderbilt Law Rev. Forthcoming, 2023, 2022

31. Cunnington, D., Law, M., Lobo, J., Russo, A.: FFNSL: feed-forward neural-symbolic learner. Mach. Learn. **112**(2), 515–569 (2023)

32. Cunnington, D., Law, M., Lobo, J., Russo, A.: Neuro-symbolic learning of answer set programs from raw data. In: International Joint Conference on Artificial Intelligence (2023)

33. Dai, W.-Z. Muggleton, S.: Abductive knowledge induction from raw data. In: Zhou, Z.-H. (ed.) Proceedings of the Thirtieth International Joint Conference on Artificial Intelligence, IJCAI-2021, pp. 1845–1851. International Joint Conferences on Artificial Intelligence Organization (2021)

34. Dai, W.-Z., Xu, Q., Yu, Y., Zhou, Z.-H.: Bridging machine learning and logical reasoning by abductive learning. In: Advances in Neural Information Processing Systems, vol. 32 (2019)

35. Dash, T., Chitlangia, S., Ahuja, A., Srinivasan, A.: A review of some techniques for inclusion of domain-knowledge into deep neural networks. Sci. Rep. **12**(1), 1040 (2022)

36. Davis, E.: Representations of Commonsense Knowledge. Morgan Kaufmann, Burlington (2014)
37. De Angelis, L., Baglivo, F., Arzilli, G., Privitera, G.P., Ferragina, P., Tozzi, A.E., Rizzo, C.: ChatGPT and the rise of large language models: the new AI-driven infodemic threat in public health. Front. Public Health **11**, 1567 (2023)
38. Dennis, L., Fisher, M., Slavkovik, M., Webster, M.: Formal verification of ethical choices in autonomous systems. Robot. Auton. Syst. **77**, 1–14 (2016)
39. Dignum, V.: Responsible Artificial Intelligence: How to Develop and Use AI in a Responsible Way. Springer Nature, Cham (2019). https://doi.org/10.1007/978-3-030-30371-6
40. Doshi-Velez, F., Kim, B.: Towards a rigorous science of interpretable machine learning. arXiv preprint arXiv:1702.08608 (2017)
41. EASA: EASA Artificial Intelligence Roadmap 1.0 A human-centric approach to AI in aviation. EASA, 02 2020
42. EASA: EASA Concept Paper: First usable guidance for Level 1 machine learning applications. EASA (2021)
43. Eisner, J., Filardo, N.W.: Dyna: extending datalog for modern AI. In: de Moor, O., Gottlob, G., Furche, T., Sellers, A. (eds.) Datalog 2.0 2010. LNCS, vol. 6702, pp. 181–220. Springer, Heidelberg (2011). https://doi.org/10.1007/978-3-642-24206-9_11
44. Ensan, A., Ternovska, E.: Modular systems with preferences. In: IJCAI, pp. 2940–2947 (2015)
45. Evans, R., Grefenstette, E.: Learning explanatory rules from noisy data. J. Artif. Intell. Res. **61**, 1–64 (2018)
46. Fagin, R., Halpern, J.Y., Moses, Y., Vardi, M.Y.: Reasoning About Knowledge. MIT Press, Cambridge (1995)
47. Ferrein, A., Lakemeyer, G.: Logic-based robot control in highly dynamic domains. Robot. Auton. Syst. **56**(11), 980–991 (2008)
48. Fischer, M., Balunovic, M., Drachsler-Cohen, D., Gehr, T., Zhang, C., Vechev, M.T.: DL2: training and querying neural networks with logic. In: Chaudhuri, K., Salakhutdinov, R., (eds.) Proceedings of the 36th International Conference on Machine Learning, ICML 2019, 9–15 June 2019, Long Beach, California, USA, vol. 97, Proceedings of Machine Learning Research, pp. 1931–1941. PMLR (2019)
49. Frieder, S., et al.: Mathematical capabilities of ChatGPT. arXiv preprint arXiv:2301.13867 (2023)
50. Furelos-Blanco, D., Law, M., Russo, A., Broda, K., Jonsson, A.: Induction of subgoal automata for reinforcement learning. In: Proceedings of the AAAI Conference on Artificial Intelligence, vol. 34, pp. 3890–3897 (2020)
51. Gajowniczek, K., Liang, Y., Friedman, T., Zabkowski, T., Van den Broeck, G.: Semantic and generalized entropy loss functions for semi-supervised deep learning. Entropy **22**(3), 334 (2020)
52. Garcez, A.S., Broda, K., Gabbay, D.M., et al.: Neural-Symbolic Learning Systems: Foundations and Applications. Springer Science & Business Media, Berlin (2002)
53. Giunchiglia, E., Stoian, M.C., Lukasiewicz, T.: Deep learning with logical constraints. In: Raedt, L.D. (ed.), Proceedings of the Thirty-First International Joint Conference on Artificial Intelligence, IJCAI 2022, Vienna, Austria, 23–29 July 2022, pp. 5478–5485. ijcai.org (2022)
54. Goodman, J.: Semiring parsing. Comput. Linguist. **25**(4), 573–605 (1999)
55. Gulwani, S.: Dimensions in program synthesis. In: PPDP, pp. 13–24. ACM (2010)
56. Hall, M., et al.: A systematic method to understand requirements for explainable AI (XAI) systems. In: Proceedings of the IJCAI Workshop on explainable Artificial Intelligence (XAI 2019), Macau, China, vol. 11 (2019)

57. Halpern, J.Y., Kleiman-Weiner, M.: Towards formal definitions of blameworthiness, intention, and moral responsibility. In: Proceedings of the 32nd AAAI Conference on Artificial Intelligence, pp. 1853–1860 (2018)
58. Halpern, J.Y., Moses, Y.: Knowledge and common knowledge in a distributed environment. J. ACM **37**(3), 549–587 (1990)
59. Halpern, J.Y., Pass, R., Raman, V.: An epistemic characterization of zero knowledge. In: TARK, pp. 156–165 (2009)
60. Hitzler, P., Sarker, M.: Neuro-symbolic artificial intelligence: The state of the art (2022)
61. Hoernle, N., Karampatsis, R.M., Belle, V., Gal, K.: Multiplexnet: towards fully satisfied logical constraints in neural networks. In: Proceedings of the AAAI Conference on Artificial Intelligence, pp. 5700–5709 (2022)
62. Huang, X., et al.: A survey of safety and trustworthiness of deep neural networks: verification, testing, adversarial attack and defence, and interpretability. Comput. Sci. Rev. **37**, 100270 (2020)
63. Huang, X., Kwiatkowska, M., Wang, S., Wu, M.: Safety verification of deep neural networks. In: Majumdar, R., Kunčak, V. (eds.) CAV 2017. LNCS, vol. 10426, pp. 3–29. Springer, Cham (2017). https://doi.org/10.1007/978-3-319-63387-9_1
64. Huang, Y.-X., Dai, W.-Z., Cai, L.-W., Muggleton, S., Jiang, Y.: Fast abductive learning by similarity-based consistency optimization. In: Advances in Neural Information Processing Systems, vol. 34 (2021)
65. Icarte, R.T., Klassen, T.Q., Valenzano, R., McIlraith, S.A.: Reward machines: exploiting reward function structure in reinforcement learning. J. Artif. Intell. Res. **73**, 173–208 (2022)
66. Ivanov, R., Weimer, J., Alur, R., Pappas, G.J., Lee, I.: Verisig: verifying safety properties of hybrid systems with neural network controllers. In: Ozay, N., Prabhakar, P., (eds.) Proceedings of the 22nd ACM International Conference on Hybrid Systems: Computation and Control, HSCC 2019, Montreal, QC, Canada, April 16-18, 2019, pp. 169–178. ACM (2019)
67. Kambhampati, S.: Challenges of human-aware AI systems. AI Mag. **41**(3), 3–17 (2020)
68. Katz, G., Barrett, C., Dill, D.L., Julian, K., Kochenderfer, M.J.: Reluplex: an efficient SMT solver for verifying deep neural networks. In: Majumdar, R., Kunčak, V. (eds.) CAV 2017. LNCS, vol. 10426, pp. 97–117. Springer, Cham (2017). https://doi.org/10.1007/978-3-319-63387-9_5
69. Katz, G., Huang, D.A., Ibeling, D., Julian, K., Lazarus, C., Lim, R., Shah, P., Thakoor, S., Wu, H., Zeljić, A., Dill, D.L., Kochenderfer, M.J., Barrett, C.: The Marabou framework for verification and analysis of deep neural networks. In: Dillig, I., Tasiran, S. (eds.) CAV 2019. LNCS, vol. 11561, pp. 443–452. Springer, Cham (2019). https://doi.org/10.1007/978-3-030-25540-4_26
70. Kleiman-Weiner, M., Saxe, R., Tenenbaum, J.B.: Learning a commonsense moral theory. Cognition **167**, 107–123 (2017)
71. Kothari, A.: ChatGPT, large language models, and generative AI as future augments of surgical cancer care. Ann. Surg. Oncol. **30**, 1–3 (2023)
72. Kwiatkowska, M., Norman, G., Parker, D.: PRISM 4.0: verification of probabilistic real-time systems. In: Gopalakrishnan, G., Qadeer, S. (eds.) CAV 2011. LNCS, vol. 6806, pp. 585–591. Springer, Heidelberg (2011). https://doi.org/10.1007/978-3-642-22110-1_47
73. Lake, B.M., Salakhutdinov, R., Tenenbaum, J.B.: Human-level concept learning through probabilistic program induction. Science **350**(6266), 1332–1338 (2015)
74. Lakemeyer, G., Levesque, H.J.: Cognitive robotics. In: Handbook of Knowledge Representation, pp. 869–886. Elsevier (2007)
75. Lemaignan, S., Ros, R., Mösenlechner, L., Alami, R., Beetz, M.: Oro, a knowledge management platform for cognitive architectures in robotics. In: IROS, pp. 3548–3553. IEEE (2010)

76. Levesque, H.J.: Common Sense, the Turing Test, and the Quest for Real AI. MIT Press, Cambridge (2017)
77. Libkin, L.: Elements of Finite Model Theory. Springer, Heidelberg (2004). https://doi.org/10.1007/978-3-662-07003-1
78. Lierler, Y., Truszczynski, M.: An abstract view on modularity in knowledge representation. In: AAAI, pp. 1532–1538 (2015)
79. Liu, C., Arnon, T., Lazarus, C., Strong, C.A., Barrett, C.W., Kochenderfer, M.J.: Algorithms for verifying deep neural networks. Found. Trends Optim. 4(3–4), 244–404 (2021)
80. Manhaeve, R., Dumancic, S., Kimmig, A., Demeester, T., De Raedt, L.: Deepproblog: neural probabilistic logic programming. In: Advances in Neural Information Processing Systems, vol. 31 (2018)
81. Marcus, G., Davis, E.: Rebooting AI: Building artificial intelligence we can trust. Vintage (2019)
82. Matarić, M.J.: The Robotics Primer. MIT Press, Cambridge (2007)
83. McIlraith, S.A., Son, T., Zeng, H.: Semantic web services. IEEE Intell. Syst. 16(2), 46–53 (2001)
84. Minervini, P., Bosnjak, M., Rocktäschel, T., Riedel, S.: Towards neural theorem proving at scale. arXiv preprint arXiv:1807.08204 (2018)
85. Minervini, P., Bošnjak, M., Rocktäschel, T., Riedel, S., Grefenstette, E.: Differentiable reasoning on large knowledge bases and natural language. In: Proceedings of the AAAI Conference on Artificial Intelligence, vol. 34, pp. 5182–5190 (2020)
86. Mitchener, L., Tuckey, D., Crosby, M., Russo, A.: Detect, understand, act: a neuro-symbolic hierarchical reinforcement learning framework. Mach. Learn. 111(4), 1523–1549 (2022)
87. Muggleton, S., et al.: ILP turns 20. Mach. Learn. 86(1), 3–23 (2012)
88. Müller, M.N., Makarchuk, G., Singh, G., Püschel, M., Vechev, M.: Prima: general and precise neural network certification via scalable convex hull approximations. Proc. ACM Program. Lang. 6(POPL), 1–33 (2022)
89. Murphy, K.: Machine Learning: A Probabilistic Perspective. The MIT Press, Cambridge (2012)
90. Müller, M.N., Fischer, M., Staab, R., Vechev, M.T.: Abstract interpretation of fixpoint iterators with applications to neural networks. In: PLDI 2023: 44nd ACM SIGPLAN International Conference on Programming Language Design and Implementation, Orlando, Florida, United States June 17–21, 2023. ACM (2023)
91. Pearl, J., Mackenzie, D.: The Book of Why. Basic Books, New York (2018)
92. Pulina, L., Tacchella, A.: An abstraction-refinement approach to verification of artificial neural networks. In: Touili, T., Cook, B., Jackson, P. (eds.) CAV 2010. LNCS, vol. 6174, pp. 243–257. Springer, Heidelberg (2010). https://doi.org/10.1007/978-3-642-14295-6_24
93. Raedt, L.D., Kersting, K., Natarajan, S., Poole, D.: Statistical relational artificial intelligence: logic, probability, and computation. Synth. Lect. Artif. Intell. Mach. Learn. 10(2), 1–189 (2016)
94. Raedt, L.D., Kimmig, A., Toivonen, H.: Problog: a probabilistic prolog and its application in link discovery. In: Proceedings of IJCAI, pp. 2462–2467 (2007)
95. Riegel, R., et al.: Logical neural networks. arXiv preprint arXiv:2006.13155 (2020)
96. Rudin, C.: Stop explaining black box machine learning models for high stakes decisions and use interpretable models instead. Nat. Mach. Intell. 1(5), 206–215 (2019)
97. Sanner, S., Kersting, K.: Symbolic dynamic programming for first-order pomdps. In: Proceedings of AAAI, pp. 1140–1146 (2010)
98. Sap, M., Shwartz, V., Bosselut, A., Choi, Y., Roth, D.: Commonsense reasoning for natural language processing. In: Proceedings of the 58th Annual Meeting of the Association for Computational Linguistics: Tutorial Abstracts, pp. 27–33 (2020)

99. Sarker, M.K., Zhou, L., Eberhart, A., Hitzler, P.: Neuro-symbolic artificial intelligence: Current trends. arXiv preprint arXiv:2105.05330 (2021)
100. Selsam, D., Lamm, M., Bünz, B., Liang, P., de Moura, L., Dill, D.L.: Learning a sat solver from single-bit supervision. arXiv preprint arXiv:1802.03685 (2018)
101. Shih, A., Darwiche, A., Choi, A.: Verifying binarized neural networks by Angluin-style learning. In: Janota, M., Lynce, I. (eds.) SAT 2019. LNCS, vol. 11628, pp. 354–370. Springer, Cham (2019). https://doi.org/10.1007/978-3-030-24258-9_25
102. Shvo, M., Klassen, T.Q., McIlraith, S.A.: Towards the role of theory of mind in explanation. In: Calvaresi, D., Najjar, A., Winikoff, M., Främling, K. (eds.) EXTRAAMAS 2020. LNCS (LNAI), vol. 12175, pp. 75–93. Springer, Cham (2020). https://doi.org/10.1007/978-3-030-51924-7_5
103. Silva, J.M., Gerspacher, T., Cooper, M., Ignatiev, A., Narodytska, N.: Explanations for monotonic classifiers. In: 38th International Conference on Machine Learning (ICML 2021), vol. 139. Machine Learning Research Press (2021)
104. Singh, G., Gehr, T., Püschel, M., Vechev, M.: An abstract domain for certifying neural networks. Proc. ACM Program. Lang. 3(POPL), 1–30 (2019)
105. Slusarz, N., Komendantskaya, E., Daggitt, M.L., Stewart, R.J., Stark, K.: Logic of differentiable logics: Towards a uniform semantics of DL. In: LPAR-24: The International Conference on Logic for Programming, Artificial Intelligence and Reasoning (2023)
106. Speichert, S., Belle, V.: Learning probabilistic logic programs over continuous data. In: Kazakov, D., Erten, C. (eds.) ILP 2019. LNCS (LNAI), vol. 11770, pp. 129–144. Springer, Cham (2020). https://doi.org/10.1007/978-3-030-49210-6_11
107. Stewart, R., Ermon, S.: Label-free supervision of neural networks with physics and domain knowledge. In: Proceedings of the AAAI Conference on Artificial Intelligence, vol. 31 (2017)
108. Szegedy, C., et al.: Intriguing properties of neural networks. In: International Conference on Learning Representations (2013)
109. Tellex, S., et al.: Approaching the symbol grounding problem with probabilistic graphical models. AI Mag. 32(4), 64–76 (2011)
110. Thielscher, M.: Reasoning Robots: the Art and Science of Programming Robotic Agents. Applied Logic Series. Springer, Dordrecht (2005). https://doi.org/10.1007/1-4020-3069-X
111. Toro Icarte, R., Klassen, T.Q., Valenzano, R.A., McIlraith, S.A.: Using reward machines for high-level task specification and decomposition in reinforcement learning. In: ICML, pp. 2112–2121 (2018)
112. Tsamoura, E., Hospedales, T., Michael, L.: Neural-symbolic integration: a compositional perspective. Proc. AAAI Conf. Artif. Intell. 35, 5051–5060 (2021)
113. Tsamoura, E., Hospedales, T., Michael, L.: Neural-symbolic integration: a compositional perspective. Proc. AAAI Conf. Artif. Intell. 35(6), 5051–5060 (2021)
114. Valmeekam, K., Olmo, A., Sreedharan, S., Kambhampati, S.: Large language models still can't plan (a benchmark for LLMS on planning and reasoning about change). arXiv preprint arXiv:2206.10498 (2022)
115. van Krieken, E., Acar, E., van Harmelen, F.: Analyzing differentiable fuzzy logic operators. Artif. Intell. 302, 103602 (2022)
116. Vaswani, A., et al.: Attention is all you need. In: Advances in Neural Information Processing Systems, vol. 30 (2017)
117. Wang, S., et al.: Beta-CROWN: efficient bound propagation with per-neuron split constraints for complete and incomplete neural network verification. In: Advances in Neural Information Processing Systems, vol. 34 (2021)
118. Williams, M.-A.: Robot social intelligence. In: ICSR, pp. 45–55 (2012)

119. Winfield, A., Blum, C., Liu, W.: Towards an ethical robot: internal models, consequences and ethical action selection. In: Proceedings of the 14th Conference Towards Autonomous Robotic Systems, pp. 85–96 (2014)
120. Winters, T., Marra, G., Manhaeve, R., Raedt, L.D.: Deepstochlog: neural stochastic logic programming. In: Thirty-Sixth AAAI Conference on Artificial Intelligence, AAAI 2022, Thirty-Fourth Conference on Innovative Applications of Artificial Intelligence, IAAI 2022, The Twelveth Symposium on Educational Advances in Artificial Intelligence, EAAI 2022 Virtual Event, February 22–March 1, 2022, pp. 10090–10100. AAAI Press (2022)
121. Wolfram, V.: Wolfram| alpha as the way to bring computational knowledge superpowers to ChatGPT. In: Stephen Wolfram Writings RSS, Stephen Wolfram, LLC, vol. 9 (2023)
122. Wooldridge, M.: An Introduction to Multiagent Systems, 2nd edn. Wiley, Chichester, UK (2009)
123. Xu, J., Zhang, Z., Friedman, T., Liang, Y., Broeck, G.: A semantic loss function for deep learning with symbolic knowledge. In: International Conference on Machine Learning, pap. 5502–5511. PMLR (2018)
124. Xu, J., Zhang, Z., Friedman, T., Liang, Y., den Broeck, G.V.: A semantic loss function for deep learning with symbolic knowledge. In: Dy, J.G., Krause, A., (eds.) Proceedings of the 35th International Conference on Machine Learning, ICML 2018, Stockholmsmässan, Stockholm, Sweden, July 10–15, 2018, vol. 80, Proceedings of Machine Learning Research, pp. 5498–5507. PMLR (2018)
125. Yang, Z., Ishay, A., Lee, J.: Neurasp: embracing neural networks into answer set programming. In: 29th International Joint Conference on Artificial Intelligence (IJCAI 2020) (2020)
126. Yang, Z., Ishay, A., Lee, J.: Neurasp: embracing neural networks into answer set programming. In: Bessiere, C., (ed.) Proceedings of the Twenty-Ninth International Joint Conference on Artificial Intelligence, IJCAI-2020, pp. 1755–1762. International Joint Conferences on Artificial Intelligence Organization (2020)
127. Yu, H., Yu, X., Lim, S.F., Lin, J., Shen, Z., Miao, C.: A multi-agent game for studying human decision-making. In: Proceedings of the 13th International Conference on Autonomous Agents and Multiagent Systems, pp. 1661–1662 (2014)
128. Zhang, H., et al.: General cutting planes for bound-propagation-based neural network verification. In: Advances in Neural Information Processing Systems (2022)

Novelty Accommodating Multi-agent Planning in High Fidelity Simulated Open World

James Chao[1]([✉]) [iD], Wiktor Piotrowski[2] [iD], Mitch Manzanares[1] [iD],
and Douglas S. Lange[1] [iD]

[1] Naval Information Warfare Center Pacific, San Diego, CA 92108, USA
{james.chao.civ,mitch.c.manzanares.civ,douglas.s.lange2.civ}@us.navy.mil
[2] Palo Alto Research Center, part of SRI International, Palo Alto, CA 94304, USA
wiktor.piotrowski@sri.com

Abstract. Autonomous agents acting in real-world environments often
need to reason with unknown novelties interfering with their plan execution. Novelty is an unexpected phenomenon that can alter the core
characteristics, composition, and dynamics of the environment. Novelty
can occur at any time in any sufficiently complex environment without any prior notice or explanation. Previous studies show that novelty
has catastrophic impact on agent performance. Intelligent agents reason with an internal model of the world to understand the intricacies
of their environment and to successfully execute their plans. The introduction of novelty into the environment usually renders their internal
model inaccurate and the generated plans no longer applicable. Novelty is particularly prevalent in the real world where domain-specific and
even predicted novelty-specific approaches are used to mitigate the novelty's impact. In this work, we demonstrate that a domain-independent
AI agent designed to detect, characterize, and accommodate novelty in
smaller-scope physics-based games such as Angry Birds and Cartpole can
be adapted to successfully perform and reason with novelty in realistic
high-fidelity simulator of the military domain.

Keywords: Novelty · Open World · Planning

1 Introduction

Current artificial intelligence (AI) systems excel in narrow-scoped closed worlds
such as board games and image classification. However, AI systems performance
drops when accommodating to constantly changing conditions [7]. On the other
hand, automated Planning has long been utilized for military applications. For
example, aircrew decision aiding modern military air missions [18], generating
complex battle plans for military tactical forces [15], handling crisis and disaster
relief [26], and controlling autonomous unmanned aerial vehicle in beyond-visual-
range combat [9]. The military not only requires a quick and decisive course of

© The Author(s), under exclusive license to Springer Nature Switzerland AG 2024
F. Amigoni and A. Sinha (Eds.): AAMAS 2023 Workshops, LNAI 14456, pp. 201–216, 2024.
https://doi.org/10.1007/978-3-031-56255-6_11

action (COA), but also flexibility to handle unforeseen situations. Good crisis management is characterized by quick response, decisive action, and flexibility to adapt to changing environments [26].

In another example, unmanned aerial vehicles (UAVs) engaged in air combat, seen in [9], continuously monitoring the actions of opponent aircraft and performing behavior recognition to predict their opponents' current plans and targets. However, this does not include enemy behavior or weapons that are far out of scope, rendering the original domain knowledge no longer feasible. Furthermore, a simple goal change or replanning can no longer resolve the problem. In order to accommodate meaningful and impactful real-world novelty, we propose integrating novelty reasoning approaches into conventional state of the art AI systems using a realistic military simulator. This is a step towards showing how AI adapts to open-real-world messiness.

In the realistic military simulator we use, there is an existing baseline AI agent that does not perform well when encountering novelty during a mission. The non-novelty agent determines its COAs using a branch-and-bound search algorithm. To improve the existing AI and be able to handle novelty will include updating the existing agent to use PDDL+ with model manipulation operators (MMOs) to monitor and repair in an event of novelty detection via a model consistency checker. Finally, an execution engine is required to translate the PDDL+ plan to run in the realistic military simulator.

In this work, we adapt Hydra [24], an existing single-agent AI planning-based novelty-aware approach to executing complex real-world scenarios in a high-fidelity military simulator. Previously, Hydra has mostly been applied to physics-based single-agent lower fidelity games such as Angry Birds [11] and OpenAI Gym's Cartpole [4]. In a high-level overview, instead of relying solely on the AI planner integrated in the simulator, we delegate the mid- and high-level decision making to Hydra while continuing to exploit the integrated AI for low-level planning and execution. The resulting composite agent architecture enables reasoning with novelty introduced in the multi-agent high-fidelity simulator which was previously not possible. The novelty-aware functionality is of great importance in military applications where environmental phenomena and enemy behavior must be accurately captured and reasoned with. We demonstrate improvement over the baseline AI agent decision making.

2   Related Work

Novelty [1,3,5,16] that is unknown, unexpected, or out of distribution is important to consider when moving AI agents from closed world games and simulations to the open real world. There exists many domains where research is currently being performed, including Monopoly [14], Minecraft [19,20], Angry Birds [11], Doom [13], Cartpole [3], natural language processing [17], and computer vision [12]. In general, novelty that does not effect any outcome of the system is considered nuisance novelty and does not have to be addressed.

Hypothesis-Guided Model Revision over Multiple Aligned Representations (Hydra) [24] is an AI framework that uses a model-based planning approach to detect, characterize, and accommodate to various novelties in multiple domains including Angry Birds, Minecraft, and Cartpole. Hydra uses PDDL+ [10], a standardized planning modeling language for mixed discrete-continuous systems, to capture the composition and dynamics of the modeled scenario. PDDL+ is a highly expressive language which enables accurate capture of dynamical system characteristics. Currently, Hydra auto-generates PDDL+ problem instance files from the environment perception information, which are then combined with a manually written domain (describing the generic system dynamics) to create a full planning model (note that the domain only needs to be written once per environment). Hydra then uses the PDDL+ planning model to detect novelty via consistency checking. A consistency checker detects divergence between the expected behavior (i.e., the planning trace) and the observations collected when executing the plan in the simulation environment. Consistency checkers can be general (i.e., taking into account all components of the environment) or focused (i.e., only focusing on a smaller subset of the environment features (e.g., unexpected behavior of a particular agent acting within the environment, or unexpected effect of an executed action). Initially, the PDDL+ planning model only accounts for non-novel behavior since novelty existence is not proved and the type of novelty is not known. Once an impactful novelty is detected via consistency checking, Hydra engages the repair module to find an explanation for the divergent behavior. The PDDL+ planning model is modified until the planning-based predictions are re-aligned with the observations. Currently, this is performed via a heuristic search process which adjusts the values of state variables such that the inconsistency (i.e., the euclidean distance between expected and observed state trajectories) is minimized.

AFSIM [8] provides a realistic simulation of behavior for the entities including F-35 fighter jets and surface-to-air missiles (SAM). The environment simulates real world environmental characteristics, including partial observability, stochasticity, multi-agent, dynamic, sequential, continuous, and asymmetric battles. An example of each characteristic is provided, but not limited to, the below examples:

- Partially observable: Nothing is observed at the beginning of a battle, the agent can use sensors to provide observations from its limited abilities.
- Stochasticity: Missiles fired at a target may not always hit, radar and sensors may not always function properly. Starting positions of battles will vary (it is worth noting that stochasticity is different from novelty that we accommodate in this paper).
- Adversarial: inherently competitive multi-agent environment.
- Dynamic: the environment is constantly changing in real time.
- Sequential: agent actions will have a long term effect for the mission at a later time. Continuous: States, time, actions are continuous.
- Asymmetric: rewards and objectives are different for the agents on opposite sides.
- Noisy: Sensor errors can cause confusion to the AI.

3   Problem Setup

We will simulate the use case from the Hollywood movie Top Gun:Maverick. Where two aircraft must cooperate and fire separate missiles to destroy the enemy target. During the mission, we consider the novelty which extends the effective range of enemy SAM missiles. This violates the assumption of the friendly force's perception and default planning models which underestimate the safe distance from the enemy SAM sites. As a result, the protagonist team led by Maverick is shot down by unexpected enemy weapons because they flew too close to the enemy, assuming they were at a safe distance.

To allow PDDL+ planning, required by the Hydra architecture, in the AFSIM environment, the environment is modified to resemble an OpenAI Gym interface [4] through a Python framework, where real time is segmented into discrete steps, missions are segmented into episodes (battles in military terms), and multiple episodes are segmented into a tournament (campaign in military terms).

Following the theory of Occam's razor [23], the states and actions will be limited to the ones relevant to the current mission using domain knowledge. Primitive actions will be combined into higher level actions. For example, actions such as tasking and routing a friendly surveillance and reconnaissance autonomous drone to gain information on the operation area will be automatically performed at the beginning of a battle.

The observations of enemy entities for both experiments will be assumed to be sensed by the surveillance drone in the beginning of all battles. There will be 10 enemy radars acting as enemy sensors (used by different military commands in the enemy chain of command), a supply depot, an ammo storage station, a SAM site, a chemical storage unit, a command post, a defense headquarters. The mission area and entities are shown in Fig. 2. The teal circle represents the starting location of the F-35 aircraft, it also represents the starting location of the autonomous surveillance drone, the red rectangle with a teal outline represents the enemy SAM location, the red circle with a teal outline represents where the target radar station is located, the red triangles denote enemy radar sensors, and the red pentagons represent all other enemy entities.

4   Planning Domain Formalization

The planning domain requires a transition system such as a simulation environment to generate a plan while maximizing a utility function such as shortest path or highest reward. Because novelty detection, characterization, and accommodation is the main research, we also require a formal definition for the novelty response problem. Following the definition of a planning domain and a concept of novelty, we use the Hydra methodology to calculate the degree in which the planning model is consistent with the environment (for our case a open-real-world-like high fidelity simulator), and finally we define a set of MMOs to facilitate a meta model repair until the model becomes more consistent with the environment. Formally:

Definition 1. *Environment Let E be the environment, a transition system defined as:*

$$E = \langle \mathcal{S}, \mathcal{S_I}, \mathcal{A}, \mathcal{E}, \mathcal{G}, \mathcal{V} \rangle \tag{1}$$

where $\mathcal{S}$ is the infinite set of states, $\mathcal{S_I} \subseteq \mathcal{S}$ is a set of possible initial states, $\mathcal{A}$ is a set of possible actions, $\mathcal{E}$ is a set of possible events, $\mathcal{G}$ is a set of possible goals, and $\mathcal{V}$ is a set of domain variables including both discrete and numeric.

Definition 2. *Novelty response problem Following [24], let $\prod$ be the novelty response problem defined as:*

$$\prod = \langle E, \varphi, t_N \rangle \tag{2}$$

where E is the transition system environment, φ is the novelty function, and t_N is a non-negative integer specifying the battle in which novelty φ is introduced within a tournament. In the military domain we operate in, the terms battle is interchangeable with episode, and campaign is interchangeable with tournament.

Definition 3. *Consistency Let C be the consistency checking function defined as:*

$$C : \mathcal{M} \times E \times \pi \times t_e \times t_o \times \mathcal{T} \to \mathbb{R}_{\geq 0} \tag{3}$$

where $\mathcal{M}$ is the PDDL+ model of mapping E to the realistic high fidelity simulator, π is a sequential plan that solves E to reach $\mathcal{G}$, t_e is the set of trajectory [22] we expect to observe using model $\mathcal{M}$ to solve the planning problem, t_o is the set of trajectory we actually observe in the environment when applying the plan π, and $\mathcal{T}$ is a threshold, where if the distance between t_e and t_o exceeds, we can assume novelty is detected. The threshold is a hyperparameter that can be "consistency shaped" based on domain knowledge. In theory, consistency checking does not required domain knowledge, in practice, it requires domain knowledge to reduce computation requirements. The domain knowledge does not map one to one to unknown novelty, but rather to parameters an agent has at its disposal to detect and accommodate the novelty.

A Model Manipulation Operator (MMO) is a single change to the agent's internal model. In this work we limit the scope of MMOs to modifying the values of variables present in the agent's internal model by a predetermined interval. MMOs are then used in the model repair mechanism to accommodate novelty by adapting the agent's internal model. In fact, repair finds a sequence of MMOs that, when applied to the agent's internal model $\mathcal{M}$ yields an updated model $\mathcal{M}'$ which accounts for the introduced novelty.

Definition 4. *MMO In the scope of this work, an MMO is a function $m :$ $V \times \Delta V \to V$, where $V \in \mathbb{R}$ is a numeric variable present in the agent's model $\mathcal{M}$, and $\Delta V \in \mathbb{R}$ is the numeric change to the value by some interval.*

In practice, an MMO is applied in a straightforward manner $v(\mathcal{M}) = v(\mathcal{M}) \pm \Delta v(\mathcal{M})$ where $v(\mathcal{M})$ is the numeric value of some variable in model $\mathcal{M}$ and

$\Delta v(\mathcal{M})$ is a predefined change interval specific to that model variable $v(\mathcal{M})$. By taking advantage of notation, this approach can also be extended to propositions $p(\mathcal{M})$ by casting each one as a numeric variable $v(p(\mathcal{M}))$. After MMOs have been applied, the variable is then re-cast into a true proposition such that $p(\mathcal{M}) = True$ if $v(p(\mathcal{M})) > 0$ and $p(\mathcal{M}) = False$ otherwise.

Following from the above MMO definition, a repair R is a function which takes in a model $\mathcal{M}$ and a sequence of MMOs $\{m\}$ that modify the model's variables. The repair returns the modified model $\mathcal{M}'$.

Definition 5. *Repair Let repair R be a function $R(\mathcal{M}, \{m\}) \to \mathcal{M}'$, where $\mathcal{M}$ is the model of the environment E, and $\{m\}$ is the a set of MMOs defined over model $\mathcal{M}$. The repair function yields an updated model $\mathcal{M}'$ which is generated by applying $\{m\}$ to the default model $\mathcal{M}$*

5   Implementing Novelty Accommodating Agents

5.1   High Level Architecture

Figure 1 describes the components and information flow of the agent architecture. The PDDL+ model generator automatically builds a PDDL+ problem from initial observations and intrinsic assumptions about the environment. A full model $\mathcal{M}$ is created by combining the auto-generated problem with a general manually-defined PDDL+ domain. A PDDL+ planner will solve the model $\mathcal{M}$ for a plan π, an execution engine will translate the plan into instructions executable in the AFSIM environment. After each battle t_N in AFSIM, the agent will calculate a consistency score C comparing the expected outcomes t_e and observed outcomes t_o. If consistency C exceeds threshold T it is likely that the underlying environment E has been substantially altered by novelty, beyond interference from noisy sensor readings. The agent then starts the meta model repair process which adjusts model $\mathcal{M}$ by iteratively applying MMOs m.

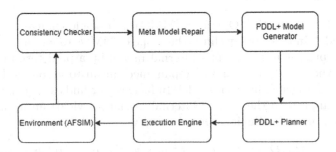

Fig. 1. Agent Architecture

5.2   PDDL+ and Execution Engine

In order to utilize the Hydra novelty detection, characterization, and accommodation, it is required to define a PDDL+ model and search for a plan to execute in the pre-novelty environment E.

The continuous mission area space is discretized into a grid space as shown in Fig. 2. The infinitely large and continuous action space is discretized into five actions: move in four directions one cell and fire the JDAM (i.e., missiles with additional precision guidance kit) when you are within range of 1 cell (28,000 m JDAM range [2], while a cell length is estimated at 25,000 m). The actions can be further broken down into similar lower level actions that will be used to execute actions in AFSIM. There will be a PDDL+ event of the F-35 aircraft being shot down by the enemy SAM if it flies within range of the enemy SAM.

Fig. 2. Mission Area Grid And Entities: teal circle is F-35 aircraft and autonomous surveillance drone, red rectangle with teal outline is the enemy SAM, red circle with teal outline is the target radar station, red triangles are enemy radar sensors, and red pentagons are all other enemy entities (Color figure online)

States, time, actions are all continuous variables in AFSIM: The continuous actions controlling air movement with high fidelity physics are taken in a continuous timeline. Time is continuous and there is no concept of a discrete time tick. Space is continuous and there is no concept of a discrete grid cell. However, the PDDL+ planner uses a discretization-based approach to solve planning tasks, discretizing it into a geo-spatial grid, time step, and discrete actions. The execution engine will use a low-level planner to translate the discrete variables used by the PDDL+ model back into continuous variables for the environment.

A move command is in one of the four cardinal directions (north, south, west, and east). Each grid cell will have its location defined as an integer $I \subseteq \mathbb{N}$ for the column and row coordinates in the grid. The engagement area composed of grid cells is irregularly shaped, meaning not a rectangular shape. The centroid point of each grid cell is mapped to a geographical latitude and longitude, and the execution engine will calculate the geo-coordinate and route the F-35 between cells. The F-35 position will be kept track using a column and row integer $I \subseteq \mathbb{N}$.

Transitioning actions between the continuous time in the real-world-like high fidelity environment and the discrete time in our PDDL+ planner provides an engineering challenge. To solve this problem, we 1) take the continuous timeline and divide it into discrete time ticks, which are chosen as a hyperparameter of the PDDL+ planner (denoted as Δt).

The domain contains move actions, which can route each aircraft between cells in cardinal directions (north, south, east, and west). A fire weapon action is also available for the agent, with the precondition that the friendly plane is in range of its target. The maximum range to fire a specific weapon is calculated using Manhattan distance between the grid cells:

$$|p_{row} - t_{row}| + |p_{col} - t_{col}| \leq p_r \tag{4}$$

Where p_{col} is the column number of the F-35 aircraft, t_{col} is the column number of the target, p_{row} is the row number of the F-35 aircraft, t_{row} is the row number of the target, and p_r is the range of the JDAM missile.

Similarly, an event is defined where the friendly plane will be shot down if it is in range of the enemy SAM missile range. Note this is using our internal model of the enemy weapon range. A PDDL+ event will have a precondition where the friendly plane will be shot down if we move within the enemy SAM range represented in Manhattan distance, with equation:

$$|s_{col} - p_{col}| + |s_{row} - p_{row}| \leq s_r \tag{5}$$

where p_{col} is the column number of the F-35 aircraft, s_{col} is the column number of the enemy SAM, p_{row} is the row number of the F-35 aircraft, s_{row} is the row number of the enemy SAM, and s_r is the range of the enemy SAM. The enemy missile range will be a numeric MMO where our agent can adjust the internal model when dealing with novelty.

The Manhattan distance (w.r.t. grid cells) of the friendly plane range is $p_r = 1$ and enemy SAM range is $s_r = 2$ is shown in Fig. 3.

Because the continuous state space is translated into a discrete grid, the enemy SAM is likely not in the exact center of a cell, therefore, the range may include additional or fewer cells than a perfect circle, the model with the longest possible range of the enemy SAM and shortest possible range of our F-35 aircraft will be used to err on the side of caution.

A scheduler will parse the plan involving every action of every agent, and convert each routing end point location to a geo-coordinate with latitude and longitude values. The center point of each grid cell, which we can get information from the environment, is mapped to geo-coordinates in the AFSIM environment, and we convert the PDDL+ plan into a AFSIM plan that can be executed in sequential order according to the PDDL+ plan. For example, a particular grid cell can be queried from the environment to map to geographic coordinates (60.14405310652675, 169.73387958469417). This data will be encoded into a pre-existing table of domain knowledge.

We introduce a hyperparameter t which designates the time delay to execute actions between aircraft. We set t to 18 s. 5 nautical miles generally allows plenty of room and distance to maneuver, for example, an evasive maneuver if an emergency was to happen, and not cause an air to air collision between the leader and the wingman. The aircraft speed is about 536 m per second [25], and 5 nautical miles is 9260 m, thus $9260m/536\frac{m}{s} = 18$ s.

Details of the PDDL+ implementation, scheduling, and execution can be found at [6].

5.3   Novelty Injection

We ran campaigns experiments $\prod$ consisting of 20 battles each. The novelty φ was injected into $\prod$ in the second battle ($t_N = 2$) for both domains. The introduced novelty are changes to the value of the enemy SAM range, from 40,000 m to 90,000 m. The expected result is the SAM range increasing from 2 cells up to 4 cells (depending on the SAM exact location), as each cell length is estimated to be 25,000 m. Further novelty examples and injections can be found in the novelty paper [5]. And the detection and accommodation to such paper will be addressed in future work.

5.4   Consistency Checking

The consistency checking $C \in \mathbb{R}_{\geq 0}$ is a calculated number representing how accurate the model $\mathcal{M}$ represents the realistic high fidelity simulator transition system E. Consistency $C = 0$ means that the agent is operating on an internal model $\mathcal{M}$ that is perfectly aligned with the simulation environment E. The higher the value of C, the less accurate the agent internal model represents the environment E. t_e is simulating using the PDDL+ model $\mathcal{M}$. t_o is generated by converting direct observations of the environment into PDDL+ to calculate C using Eq. 6 (because t_e and t_o has to be the same datatype). While comparing every single variable V will be ideal in catching any model inconsistency, however, computation limitations of calculating ∞ amounts of ΔV makes that infeasible, so we focus on comparing the MMOs and the Euclidean distance between the trajectories of the defined MMOs that are defined using domain knowledge.

$$C = \sum_i \gamma^i \cdot \|t_o(\pi, E)[i] - t_e(\pi, \mathcal{M})[i]\| \tag{6}$$

where $t_o(\pi, E)$ is the observed trajectory of executing the plan π in the environment E. And $t_e(\pi, \mathcal{M})$ is the expected trajectory of executing the plan π using the PDDL+ model. $0 < \gamma < 1$ is a discount factor to account for compounding errors of Euclidean distances for later states.

5.5   Meta Model Repair

The meta model repair is triggered once the consistency C of the default model $\mathcal{M}$ exceeds the threshold $\mathcal{T}$. Repair aims to alter the model $\mathcal{M}$ so that the expected trajectory t_e (yielded by the planner, based on $\mathcal{M}$) is consistent with the trajectory t_o observed when executing plan π in simulation environment E. As stated in [24], defining appropriate MMOs is the key to a good repair that leads to good novelty accommodation. Although monitoring every variable $v \in \mathcal{M}$ will create the most general agent, the search space might become too large to find feasible repairs in a reasonable time. Thus, currently, model variables which MMOs will modify during repair are chosen manually. In the presented missions, as a proof of concept, we labeled the missile range variable as "repairable" via MMOs being one of the most mission critical variables V in our environment E. The main observation of interest is whether the F-35 aircraft and the enemy target is destroyed. Consistency score will yield $C = 1.0$ if the F-35 aircraft is destroyed and the enemy target is fully functional, and a $C = 0.0$ score if the F-35 aircraft survives and the enemy target asset is destroyed.

Algorithm 1: PDDL+ meta model repair algorithm

$C_{best} \leftarrow \text{EstimateConsistency}(\mathcal{M}, \pi, t_e, t_o, \mathcal{T})$
while $C_{best} \geq \mathcal{T}$ **do**
    forall the $MMO \in m$ **do**
        $\mathcal{M}' \leftarrow R(\mathcal{M}, MMO)$ $C_{new} \leftarrow \text{EstimateConsistency}(\mathcal{M}, \text{E}, \pi, t_e, t_o, \mathcal{T})$
        if $C_{new} < C_{best}$ **then**
            $C_{best} \leftarrow C_{new}$ **if** $C_{best} \leq \mathcal{T}$ **then**
                | $\mathcal{M} \leftarrow \mathcal{M}'$
            end
    end
    $\text{UndoUpdate}\mathcal{M}(MMO)$
    end
    return $\mathcal{M}$
end

The search-based algorithm for model repair is domain-independent. The model repair will follow Algorithm 1, which shows the pseudo-code for the meta model repair. First, we check the consistency score C_{best}, if $C_{best} > \mathcal{T}$ then we repair the MMOs {m} one by one. For each repaired MMO, we check the consistency C_{new}, if any C_{new} is smaller than the best consistency score C_{best}, then we deem the MMO repair as successful and update the model $\mathcal{M}$ until we find the smallest C_{best} possible, which reflects the least amount of model inconsistency.

The procedure turns this problem into a search task which iteratively considers different changes to the model via predefined MMOs (e.g., increase range by 1 cell, increase range by 2 cells, decrease range by 1 cell, etc.). The goal of this search is to sufficiently reduce the inconsistency between expected and observed state trajectories. The procedure is adapted from Stern et al., 2022, "Model-Based Adaptation to Novelty for Open-World AI".

6   Experiment Setup

Experimental evaluation was conducted on a machine with MacOS, an Intel Core i7 2.6 GHz 6 core processor with 16 GB 2667 MHz DDR4 memory.

Since the baseline agents have no concept of novelty, detection and false positives are both 0%. And the mission is a total failure 37% of the time, meaning the F-35 is destroyed, and the target is still fully functional. An interesting result is that the win percentage is relatively high, succeeding in the mission over half of the tries. This leaves an interesting final decision whether to execute the mission in the next battle or not, knowing the mission can either be a successful strike or end in total failure, without any reasonable explanation.

6.1   Novelties

In this mission a F-35 aircraft is given a task to strike a radar station device roughly 90,000 m south-west of a SAM missile launcher in enemy territory. A plan is generated using a model of the world given to it before the mission begins. In the scenario the agent can successfully plan a strike given its prior information is correct. However, when the information it assumes to be true is incorrect, it is unable to develop a successful plan and the F-35 aircraft is destroyed by the SAM site. Figure 3 shows the pre-novelty and post-novelty range of the SAM, the dotted yellow line shows the pre-novelty range, and the dotted purple line shows the post-novelty range.

A 20 battle campaign $\prod$ with novelty φ injected in the second battle ($t_N = 2$) is ran, every battle it generates a plan using the agent internal model of the world. In the pre-novelty scenarios, the agent can successfully generate a plan to complete the mission. The agent search statistics is shown in Table 2.

After novelty injection, the SAM missile range is increased, using the old model, it is unable to develop a successful plan and the aircraft risks being destroyed by the SAM upon entering the new SAM range. Due to environment stochasticity described in the related work section, it takes the risk of entering the SAM without knowing it has and successfully executes the mission 15 times, and is destroyed 4 times post-novelty. The performance of novelty accommodation is shown in Table 1. Since the baseline agents have no concept of novelty, detection and false positives are both 0%. And the mission becomes a total failure 21% of the time, resulting in the F-35 aircraft being destroyed and the enemy target still functional. An interesting result is that the win % is very high, succeeding in the mission at a high rate. Perhaps making a human decision to execute the

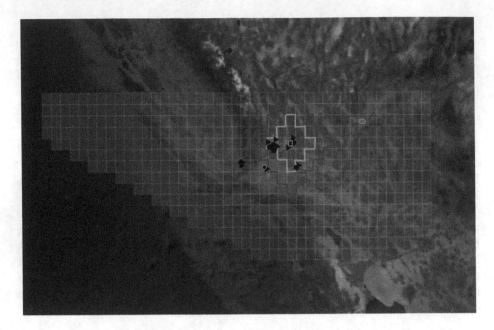

Fig. 3. SAM Range Before And After Novelty

mission difficult knowing the statistics, as 79% of the time, it results in a mission success.

7   Results

Of the 20 battle campaign $\prod$, the first battle the agent successfully completes the mission with no novelty. The second battle includes the missile range novelty, and the agent proceeds to complete the mission, the reason is due to environment stochasticity of high fidelity simulators mentioned in the related work section, the aircraft can unknowingly takes the risk of entering the SAM missile range area without knowing it has and successfully executes the mission sometimes, but is destroyed other times without knowing why. The agent performance of novelty detection and accommodation is shown in Table 1 using performance measurement described in [21].

For the fourth battle through the last, the consistency function C passes the threshold $\mathcal{T}$, the consistency score C becomes 1 because the distance between t_e and t_o is now 1 for losing the aircraft last battle, and repair is turned on. The SAM range is the numeric MMO $\{m\}$ monitored in the repair R; the numeric value of the SAM range is adjusted to increase from 2 cells to 4 cells. The MMO change ΔV resulted in decreased consistency score C and the repair is deemed successful, it is also able to characterize the novelty by repairing "ss-weapon-range" from 2 to 4. With the repair, the agent will navigate around the risky area starting from the next battle after novelty is detected. As shown in Fig. 4, the agent is able to destroy the enemy target with a new route after repairing the PDDL+ model. The yellow solid line shows the pre-novelty route, while the purple solid line denotes the post-novelty route. The grey symbol of a bomb with dotted teal outlines represents the location where the F-35 fired the JDAM missile from, in both cases the F-35 aircraft proceeds to return to home base after firing the JDAM.

Fig. 4. Agent Plan Before And After Novelty

Table 1. Performance measurement of baseline and novelty accommodating agents

	Baseline	Hydra
False neg%	21	5
Win% post-Nov	79	95
Win% pre-Nov	100	100
Detection%	0	95
False pos%	0	0

Table 2. Pre-Novelty Agent Plan

Planning Time	Plan Length	States Explored
19.118	33	2632

Table 3. Novelty Accommodating Agent Plan

Planning Time	Plan Length	States Explored
18.312	41	2421

The performance of the novelty accommodating agent is in Table 1. Of the 20 battles, compared to the baseline agent, our novelty detection false positive rate is still 0%. The false negative rate decreased to 5%. The pre-novelty win rate is 100%, and the post novelty win rate increases to 95%. In this tactical mission result, there are no complications, the agent completes the mission and is not destroyed. Novelty detection rate increases to 95%.

The novelty accommodating agent search statistics is shown in Table 3. The novelty accommodating plan explored less states but the plan is longer due to navigating around a longer range. While the low level primitive actions in AFSIM are still combined and abstracted, routing options are fixed to geo-coordinates defined as the centroids of each grid cell, and all the primitive firing actions are combined into one action for a range of one cell.

8   Conclusion and Future Work

We demonstrated a prototype framework to utilize the Hydra AI system in a realistic high fidelity simulator that is being used to execute intricate military scenarios, and novelties that resemble open-real-world problems. Our results prove that model-based AI systems like Hydra can be adapted from smaller-scoped games with hypothetical game related novelties to more realistic applications.

The next step towards accommodating real open-world novelty is to extend this framework and model a richer domain including additional real-world novelties and complexity such as hidden enemy units, faulty sensors, unknown delays, different types of weapons, different environmental effects, different capabilities by enemy fighting units, complex routing for different mission objectives, and unknown civilian obstacles. Furthermore, in future scenarios, we plan to add more actions, agents, and goal complexity to the scenario, to allow the agent to explore more complex decisions and novelty detection, characterization, and accommodation opportunities.

Acknowledgement. This research was sponsored by DARPA. The views contained in this document are those of the authors and should not be interpreted as representing the official policies, either expressed or implied, of DARPA or the U.S. government.

References

1. Alspector, J.: Representation edit distance as a measure of novelty. arXiv preprint arXiv:2111.02770 (2021)
2. Bell, W.: Joint direct attack munition (JDAM). Technical report, US Air Force Hill AFB United States (2015)
3. Boult, T., et al.: Towards a unifying framework for formal theories of novelty. In: Proceedings of the AAAI Conference on Artificial Intelligence, vol. 35(17), pp. 15047–15052 (2021)
4. Brockman, G., et al.: OpenAI gym. CoRR abs/1606.01540 (2016). arxiv.org/abs/1606.01540
5. Chadwick, T., et al.: Characterizing novelty in the military domain (2023). arxiv.org/abs/2302.12314
6. Chao, J., Piotrowski, W., Manzanares, M., Lange, D.S.: Top gun: cooperative multi-agent planning. In: AAAI Workshop on Multiagent Path Finding (2023). www.idm-lab.org/wiki/AAAI23-MAPF/index.php/Main/HomePage?action=download&upname=Paper_4.pdf
7. Chao, J., Sato, J., Lucero, C., Lange, D.S.: Evaluating reinforcement learning algorithms for evolving military games. Convergence **200**(44), 1 (2020)
8. Clive, P.D., Johnson, J.A., Moss, M.J., Zeh, J.M., Birkmire, B.M., Hodson, D.D.: Advanced framework for simulation, integration and modeling (AFSIM) (case number: 88abw-2015-2258). In: Proceedings of the International Conference on Scientific Computing (CSC), p. 73. The Steering Committee of The World Congress in Computer Science, Computer ... (2015)
9. Floyd, M.W., Karneeb, J., Moore, P., Aha, D.W.: A goal reasoning agent for controlling UAVs in beyond-visual-range air combat. In: IJCAI, vol. 17, pp. 4714–4721 (2017)
10. Fox, M., Long, D.: Modelling mixed discrete-continuous domains for planning. J. Artif. Intell. Res. **27**, 235–297 (2006)
11. Gamage, C., Pinto, V., Xue, C., Stephenson, M., Zhang, P., Renz, J.: Novelty generation framework for AI agents in angry birds style physics games. In: 2021 IEEE Conference on Games (CoG), pp. 1–8 (2021). https://doi.org/10.1109/CoG52621.2021.9619160
12. Girish, S., Suri, S., Rambhatla, S.S., Shrivastava, A.: Towards discovery and attribution of open-world GAN generated images. In: Proceedings of the IEEE/CVF International Conference on Computer Vision (ICCV), pp. 14094–14103 (2021)
13. Holder, L.: holderlb/wsu-sailon-ng (2022). www.github.com/holderlb/WSU-SAILON-NG
14. Kejriwal, M., Thomas, S.: A multi-agent simulator for generating novelty in monopoly. Simul. Model. Pract. Theory **112**, 102364 (2021). https://doi.org/10.1016/j.simpat.2021.102364, www.sciencedirect.com/science/article/pii/S1569190X21000770
15. Kewley, R.H., Embrechts, M.J.: Computational military tactical planning system. IEEE Trans. Syst. Man. Cybern. Part C. App. Rev. **32**(2), 161–171 (2002)
16. Langley, P.: Open-world learning for radically autonomous agents. In: Proceedings of the AAAI Conference on Artificial Intelligence, vol. 34(09), pp. 13539–13543 (2020)
17. Ma, N., et al.: Semantic novelty detection in natural language descriptions. In: Proceedings of the 2021 Conference on Empirical Methods in Natural Language Processing, pp. 866–882. Association for Computational Linguistics, Online and

Punta Cana, Dominican Republic, November 2021. https://doi.org/10.18653/v1/2021.emnlp-main.66, www.aclanthology.org/2021.emnlp-main.66

18. Merkel, P.A., Lehner, P.E., Geesey, R.A., Gilmer, J.B.: Automated planning with special relevance to associate systems technology and mission planning. Technical report, BDM International Inc., Arlington, VA (1991)

19. Muhammad, F., et al.: A novelty-centric agent architecture for changing worlds. In: Proceedings of the 20th International Conference on Autonomous Agents and MultiAgent Systems, pp. 925–933 (2021)

20. Musliner, D.J., Pelican, M.J., McLure, M., Johnston, S., Freedman, R.G., Knutson, C.: Openmind: planning and adapting in domains with novelty. In: Proceedings of the Ninth Annual Conference on Advances in Cognitive Systems, pp. 1–20 (2021)

21. Pinto, V., Renz, J., Xue, C., Doctor, K., Aha, D.: Measuring the performance of open-world AI systems. In: Proceedings of the AAAI conference on Designing Artificial Intelligence for Open Worlds (2020)

22. Piotrowski, W.M., Fox, M., Long, D., Magazzeni, D., Mercorio, F.: Heuristic planning for PDDL+ domains. In: Workshops at the Thirtieth AAAI Conference on Artificial Intelligence (2016)

23. Russell, S., Norvig, P.: Artificial Intelligence: A Modern Approach, Global 4th Edn. Foundations, vol. 19, 23 (2021)

24. Stern, R., et al.: Model-based adaptation to novelty for open-world AI. In: Proceedings of the ICAPS Workshop on Bridging the Gap Between AI Planning and Learning (2022)

25. Wiegand, C.: F-35 air vehicle technology overview. In: 2018 Aviation Technology, Integration, and Operations Conference, p. 3368 (2018)

26. Wilkins, D.E., Desimone, R.: Applying an AI planner to military operations planning. Citeseer (1993)

Towards Reducing School Segregation by Intervening on Transportation Networks

Dimitris Michailidis(✉), Mayesha Tasnim, Sennay Ghebreab, and Fernando P. Santos

Civic AI Lab, Socially Intelligent Artificial Systems, Informatics Institute, University of Amsterdam, Amsterdam, The Netherlands
{d.michailidis,m.tasnim,s.ghebreab,f.p.santos}@uva.nl

Abstract. Urban segregation is a complex phenomenon associated with different forms of social inequality. Segregation is reflected in parents' school preferences, especially in context of free school choice modes. Studies have shown that parents consider both distance and demographic composition when selecting schools for their children, potentially exacerbating levels of residential segregation. This raises the question of how intervening on transit networks—thereby affecting school accessibility to citizens belonging to different groups—can alleviate spatial segregation. In this work-in-progress paper, we propose a new agent-based model to explore this question. Conducting experiments in synthetic and real-life scenarios, we show that improving access to schools via transport network interventions can lead to a reduction in school segregation over time. The mathematical framework we propose provides the basis to simulate, in the future, how the dynamics of citizens preferences, school capacity and public transportation availability might contribute to patterns of residential segregation.

Keywords: Transportation Networks · One-sided Matching · Agent-based Simulations · Dynamic Preferences

1 Introduction

Urban segregation is a complex phenomenon that reverberates across multiple socio-economic contexts—from social mobility to educational opportunities. In the context of education, centralized school admissions systems such as *Deferred Acceptance* and *Random Serial Dictatorship* have been popularized across the world for their simplicity and fairness in student allocation [2,8]. However, school segregation can emerge in such preference-based systems, reflecting (or even amplifying) existing residential segregation patterns [6]. There is evidence that parents do not send their children to schools in their residential neighborhoods; if they did, schools would be less segregated than how they currently are [12].

Although parents might prefer schools outside their neighborhoods, distance and commuting time are important factors for attending a school [6]. With the

© The Author(s), under exclusive license to Springer Nature Switzerland AG 2024
F. Amigoni and A. Sinha (Eds.): AAMAS 2023 Workshops, LNAI 14456, pp. 217–227, 2024.
https://doi.org/10.1007/978-3-031-56255-6_12

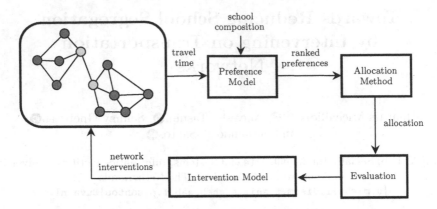

Fig. 1. Proposed agent-based model to study the impact of transport network interventions on school segregation. We consider an environment where citizens, schools, and a transportation graph are distributed in space (Sect. 2.1). At each round, agents A generate preferences for schools F, using a preference model (Sect. 2.2). Agents are assigned to schools via an allocation method (Sect. 2.3), which is evaluated on segregation (Sect. 2.4). An intervention model creates edge-based interventions to the transportation network, aiming to reduce segregation (Sect. 2.5).

exception of high-income households, most do not tend to move house and thus their choice is limited by their location [3]. Intervening on public transportation networks can thereby affect segregation, by allowing citizens from different societal groups to attend a wider set of schools. This raises a natural question: *Can transportation networks be designed, or extended, to efficiently reduce school segregation?*

Here we resort to agent-based modeling (ABM) to explore the previous question. Prior studies focused on the complexity of residential and school segregation via ABMs [6,15], and preference models based on both school composition and distance have been explored [6,14]. However, these works do not study the effect of strategically increasing accessibility to specific schools. Graph-based interventions have been utilized before to reduce accessibility inequality [10], but not to tackle school segregation. We assess whether graph-based transportation interventions can be used to reduce disparities in group composition within schools, under a centralized admission system.

We test transport network intervention strategies based on greedy optimization of classic graph centrality measures such as *closeness*, *betweenness*, and *degree* centrality. We conduct experiments in a synthetic and a real-life environment in the city of Amsterdam and show that targeted interventions can lead to a significant reduction in segregation over time (Fig. 1).

2   Methods

2.1   Environment: Citizens, Transportation, and Schools

We model the environment as an undirected graph $\mathbb{G} = (V, E)$, where $V = \{v_1, ..., v_{n_v}\}$ are nodes, one for each census tract in the city, and $E = \{e_{i,j}\}, i, j \in V, i \neq j$ are edges that represent transportation connections between nodes. For the sake of simplicity, the edges are unweighted, but the model can be used with weighted edges too (e.g., representing transportation times). We define the shortest path between i and j as $t_{i,j}$, $i, j \in V$.

We define a set of N agents (citizens), $A = \{a_1, ..., a_N\}$. An agent a is characterized by its residence node $v_a \in V$. Each agent belongs to a group $g \in G$, defined based on characteristics such as ethnicity, income, or other socioeconomic status. Finally, each agent has a homophily attribute, $h_i \in [0, 1]$, defining a preference for an optimal fraction of agents from the same group attending a school [6,11]. Note that agents are abstract entities that represent students in a city.

We define schools $f \in F$, which are located in nodes $v_f \in V$. Each school is associated with a capacity (maximum number of allowed agents) $s_f \in [0, N]$ and a group composition (fraction of assigned agents from each group) $c_{g,f} \in [0, 1]$, $g \in G$. Note that $\sum_g c_{g,f} = 1$, $\forall f \in F$.

2.2   Preference Model

At every round, each agent $a_i \in A$ creates a preference list $P_i \subseteq F$, over schools. Each school appears once on each list. The preference list is based on a utility function $U_{if}, f \in F$, and schools are sorted in descending order. We adopt the widely used Cobb-Douglas utility function, based on a function of school composition $C : c_{g,f} \to \mathbb{R}$ and travel time from the agent's residence to the school $t_{i,f}$ [6,14]

$$U_{i,f} = c_{g,f}^{\alpha} \, t_{i,f}^{(1-\alpha)}, \tag{1}$$

where g denotes the group that agent a_i belongs to and $0 \leq \alpha \leq 1$ is a parameter that controls the weight of the group composition over the travel time. Travel time is normalized by the maximum value and is calculated as follows [6]:

$$t'_{i,f} = \begin{cases} \frac{t_{max,i} - t_{i,f}}{t_{maxi} - t_{min,i}}, & \text{if } t_{i,f} \leq t_{max,i} \\ 0, & \text{otherwise.} \end{cases} \tag{2}$$

For the school composition, we use a single-peaked utility function, that is maximized when the number of agents of the same group in a school $x_{g,f}$ is equal to the homophily attribute h_i [6,14]. Values above h_i incur a constant penalty M:

$$C(x_{g,f}, h_i, M) = \begin{cases} \frac{x_{g,f}}{h_i}, & \text{if } x_{g,f} \leq h_i \\ M + \frac{(1-x_{g,f})(1-M)}{1-h_i}, & \text{if } x_{g,f} > h_i \end{cases} \tag{3}$$

M controls the level of dissatisfaction when the fraction of similar agents exceeds the optimal h_i. With this formulation interventions in the transportation network are performed to reduce the travel time $t_{i,f}$ of agents to school, with the goal of increasing utility towards more segregated schools.

2.3 Allocation Method

Once the preference lists P have been generated at each simulation round for all $a_i \in A$, they are then provided as input to an allocation method R. R is defined as a function $R : P \to F$ which takes as input a preference list p_i for agent a_i and capacity s_f for all $f \in F$ and assigns a school $f_i \in p_i$. Random Serial Dictatorship (RSD) is a popular mechanism for one-sided matching between schools and students [2]. In RSD a lottery number is first uniformly drawn for each student. The students are then serially allocated to the top-preferred school with remaining capacity in increasing order of the lottery. For our simulations we implement RSD and perform allocations at every round; schools have, overall, capacity to allocate all students, i.e., $\sum s_f \geq N$. Additionally, for each student the preference model from Sect. 2.2 provides a ranking for all schools, and RSD can allocate all students. The allocation result is then aggregated for evaluation.

2.4 Allocation Evaluation

After each simulation round, the allocation of agents to schools is evaluated on segregation. To measure segregation, we use the Dissimilarity Index (DI), a measure that captures the differences in the proportions of agents from two groups assigned to a school [7]. DI has been widely used in assessing segregation, as it takes into account the total number of agents from each group, making it suitable to use even when one group is a minority [1]. DI is defined as follows:

$$DI = \frac{1}{2} \sum_{f=1}^{|F|} |\frac{g_{1,f}}{G_1} - \frac{g_{2,f}}{G_2}|, \quad DI \in [0,1] \tag{4}$$

where $g_{j,f}$ is the number of agents of group j in school f; G_j is the number of agents in group j. Segregation is minimum (maximum) when $DI = 0$ ($DI = 1$).

2.5 Intervention Model

We explore the impact that intervening on public transport networks has on school choices. By improving transportation, we aim to elevate the rank of schools composed of majority groups in the preference lists of minority groups, increasing their accessibility to popular (yet distant) schools. Transport interventions are performed in the form of graph augmentations, by creating a new edge set $E' : \mathbb{G}, B \to \mathbb{G}'$ to the spatial graph, under a budget B [10]. It follows that $\mathbb{G}' = (V, E \cup E')$. Interventions can be seen as a proxy to the creation/expansion of public transportation lines in a real city, such as bus, metro or tram. We

(A) SBM Environment   (B) Amsterdam Environment

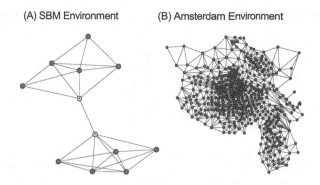

Fig. 2. We study synthetic (left image) and real (right image) environments. Nodes represent neighborhoods and yellow nodes (marked with *) indicate nodes with schools. (Color figure online)

constrain the total number of interventions to a budget B, reflecting resource limitations.

The goal of interventions is to find the best set of edges E' to add to the graph, such that total segregation is reduced. Segregation depends on the allocation method (Sect. 2.3), which has a random element to it. Therefore, optimizing directly for the dissimilarity index is not possible. We look for targeted interventions that increase accessibility to certain schools for certain groups, aiming to affect the agent's preferences in such a way that segregation is reduced.

We test two classes of greedy interventions: 1) **Centrality** and 2) **Group-based Centrality Optimization**. We identify the schools that have the lowest network centrality measure (*closeness, betweenness* or *degree*) [4] with respect to any group and then add the intervention that leads to the maximum increase in that node's corresponding 1) centrality or 2) group-based centrality.

3 Experimental Setup

We perform experiments on two graph environments: a real-life city environment based on Amsterdam neighborhoods, demographic and transportation data; and a synthethic environment based on the stochastic block model (SBM) [9], which allows us to have full control over the level of modularity and segregation in a hypothetical city. For more details please refer to Appendix B and Fig. 2.

4 Preliminary Results

In Fig. 3, we present the 95% confidence interval of the Dissimilarity Index on each simulation round, over a total of 50 rounds. Our preliminary experiments show that, under the settings outlined above, all targeted network intervention strategies proposed in Sect. 2.5 lead to a significant reduction of segregation over

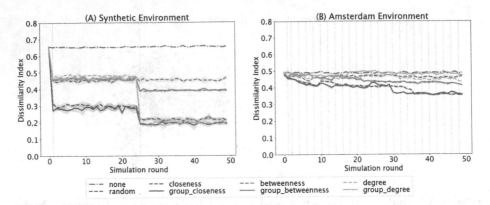

Fig. 3. We show that targeted interventions in the network can significantly decrease segregation over time. Strategies based on closeness perform best over other centrality measures. Vertical dashed lines indicate rounds with graph interventions.

time, when compared to a no-intervetion scenario (none) or random interventions. Specifically, we observe that greedy interventions aimed at increasing the closeness of the least-accessible nodes lead to the highest reduction of segregation over time. We also observe that degree-based interventions can have similar effects to closeness, but only in small networks, like SBM. This is because, when the number of nodes is low, increasing the degree of a node also increases its closeness to other nodes. A betweenness-based strategy reduces segregation and outperforms degree-based ones in a bigger environment, like that of Amsterdam. Finally, there are seemingly no significant differences between centrality and group-based centrality strategies, but depending on the budget, group-based closeness can outperform its classic counterpart.

5   Conclusion and Future Work

In this work-in-progress paper, we used an agent-based simulation model to study the impact of transport network interventions on school segregation, under the prevalence of a centralized school choice algorithm. We have demonstrated in both a synthetic and a real-life environment that, by affecting citizens preferences for particular schools, targeted transportation interventions can ultimately reduce school segregation over time. In the future, we plan to further experiment with the parameters of the preference model, to assess the sensitivity of network interventions to different types of agent school preferences. We plan to further experiment with group-based interventions, aiming at identifying the contexts where they become more efficient than centrality-based interventions.

Acknowledgements. This research was supported by the Innovation Center for AI (ICAI, The Netherlands) and the City of Amsterdam.

Appendix

A   Intervention Methods

Section 2.5 introduces the design choice to test two classes of greedy algorithms in the intervention model of the ABM. The algorithms and their usage as intervention methods are discussed below:

A.1   Greedy Centrality Optimization

Making a school more accessible is a non-trivial optimization problem, especially for large graphs [5]. We use a greedy algorithm to approximate the optimal set of interventions to apply to the graph with respect to accessibility. This translates to increasing a school node centrality $\mathbb{C}$ with respect to the other nodes. We evaluate strategies based on the classic graph measures of *closeness* ($\mathbb{C}_C$), *betweenness* ($\mathbb{C}_B$), and *degree* ($\mathbb{C}_D$) centrality.

At every intervention step, we find the school that has the lowest centrality measure with respect to any group and then add the intervention that leads to the maximum increase in this node's centrality. The process is described in Algorithm 1.

Algorithm 1. Greedy Centrality Optimization

Input $\mathbb{G} = (V, E)$

$E' \leftarrow \{\}$
for $b = 1, 2, ...B$ **do**
    $v_{g_{min}} \leftarrow argmin\{\mathbb{C}(v, g) \mid v \in V, g \in G\}$
    $\mathbb{C}_{max} = 0$
    $e_{max} \leftarrow null$
    for $u \in V$, $u \neq V$ **do**
        $e \leftarrow (u, v)$
        Compute $\mathbb{C}(v_{g_{min}}, E \cup E' \cup e)$
        if $\mathbb{C}(v_{g_{min}}, E \cup E' \cup e) > \mathbb{C}_{max}$ **then**
            $\mathbb{C}_{max} = \mathbb{C}(v_{g_{min}}, E \cup E' \cup e)$
            $e_{max} \leftarrow e$
        end if
    end for
    $E' \leftarrow E' \cup e_{max}$
end for
Output $\mathbb{G}' = (V, E \cup E')$

A.2   Group-Based Centrality

Classic centrality measures fail to capture group dynamics in a graph. In segregated environments like cities, central areas can exhibit high closeness centrality, despite having low accessibility to specific groups. Examples of this phenomenon include cities where low-income households concentrate in the outskirts, while high-income households are situated closer to the center. To account for this disparity in measurement, we introduce group-based extensions of the classic centrality measures $\mathbb{C}^g$, $g \in G$, that take into account the distribution of groups within nodes. These are namely group-based closeness $\mathbb{C}_C^g$, betweenness $\mathbb{C}_B^g$ and degree $\mathbb{C}_D^g$. Let D_g, $g \in G$ be the distribution of group g on all nodes V in the network such that $\sum_g D_g = 1$.

Group-Based Closeness Centrality. Group-based closeness $\mathbb{C}_C^g$ of a node $v \in V$ is defined as the reciprocal of the sum of travel times from all other nodes u, weighted by the fraction of agents of group g in u, $p(g|u)$.

$$\mathbb{C}_C^g(v) = \sum_u \frac{1}{t(u,v)\, p(g|u)} \tag{5}$$

where $t(u,v)$ is the travel time between nodes u and v.

Group-Based Betweenness Centrality. Group-based betweenness $\mathbb{C}_B^g$ of a node $v \in V$ is defined as the number of shortest paths σ from all nodes $o \in V$ to all nodes $d \in V, o \neq d$, that pass through v, weighted by the fraction of agents of group g in o. $p(g|o)$.

$$\mathbb{C}_B^g(v) = \sum_{o \neq v \neq d} \frac{\sigma_{t_{o,d}}(v)}{\sigma_{t_{o,d}}} p(g|o) \tag{6}$$

Group-Based Degree Centrality. Group-based degree $\mathbb{C}_D^g$ of a node $v \in V$ is defined as the total number of edges connected to a node $E_v = e_{u,v}, u \in V, u \neq v$, weighted by the fraction of agents of group g in u, $p(g|u)$.

$$\mathbb{C}_D^g(v) = \sum_{u \in V, u \neq v, e_{u,v} \in E} p(g|u) \tag{7}$$

Optimizing for group-based centrality measures leads to interventions that target schools where specific groups are underrepresented, instead of arbitrarily increasing the centrality of a school.

B   Simulation Environments

We perform experiments on two graph environments, a synthetic stochastic block model (SBM) [9] and a real city environment based on Amsterdam, Netherlands.

SBM Environment. The SBM graph is specifically generated to form clusters of communities, where nodes are densely connected with other nodes in their community and scarcely connected with nodes outside of it. We generated an SBM graph of $n_v = 12$ nodes and $n_e = 27$ edges; nodes clustered in 2 communities, which represent the majority group of their respective nodes. The parameters are chosen specifically to create a highly segregated graph, in which we aim to study the impact of the proposed intervention strategies. In-community edge probability is set to 0.7 and out-community probability is set to 0.01. In Fig. 2 (A) we show the realization of the SBM graph we used for the simulation.

Further, we generated a population of $N = 1000$ agents and sampled both their residence node and their group membership, from a total of 2 groups. Group samples are chosen in such a way that each group, within their respective community has a majority of ≥ 0.8 and outside of their community a minority of ≤ 0.2. Since agents do not start at random nodes and there is no moving action in the model, we assume that the optimal fraction of similar agents is equal to the fraction of the majority group of each node. Formally, the homophily parameter of an agent i in a node v is set to $h_{i,v} = \max\{c_{g,v}\}$, $g \in G$, where $c_{g,v}$ is the composition of group g in node v.

Finally, we place two schools on the graph, located in the two most connected nodes of the SBM graph. The initial group composition of each school is set to be equal to the group composition of the node it is located in.

Amsterdam Environment. To model the real-life environment of Amsterdam, we create a graph where census tracts are converted to nodes, which are connected with their neighboring tracts via an unweighted edge. This graph structure has recently been used to quantify segregation because it provides a scale-free and generalizable method [13]. In total, the graph consists of $n_v = 517$ nodes and $n_e = 1611$ edges. In Fig. 2 (B) we show the graph used for the Amsterdam experiments.

Similar to SBM, we generate a population of $N = 7000$ agents. However, in this environment, agents are generated to represent the real-life population of Amsterdam and are split in groups of western (W) and non-western (NW) ethnic background. More details on the population can be found in Table 1. Here the homophily parameter is set in the same way as in the SBM environment.

We use the publicly available Amsterdam secondary school dataset provided by DUO[1] which contains 47 secondary schools and their locations. We combine this information with the admissions dataset collected by OSVO[2] which provides the capacities for each school based on the admission results of the previous year.

---

[1] Education Executive Agency: http://duo.nl.
[2] The association of school boards in Amsterdam: https://www.verenigingosvo.nl/.

Table 1. Parameters used in running the experiments.

	SBM	Amsterdam
Groups	g0, g2	W, NW
Total Population	1000	7000
Group Populations	524, 476	4547, 2453
Group Populations (%)	52%, 48%	65%, 35%
No. of Nodes	12	517
No. of Edges	27	1611
α, M	0.2, 0.6	0.2, 0.6
Budget (B)	1	1
Simulation Rounds	50	50
Allocation Rounds	5	5
Interventions	2	25

B.1 Simulation Parameters

For the experiments shown in this work, we follow the setup of Dignum et al. and set the relative weight of the composition in the preference model to $\alpha = 0.2$ and the constant $M = 0.6$ for both environments. All experiments are run over 50 simulation rounds, with 5 random serial dictatorship allocations at each environment. We perform 2 intervention rounds in the SBM environment with a budget of $B = 1$ each, while in Amsterdam, we perform a total of 25 intervention rounds, also with $B = 1$. Parameters including total number of intervention rounds and budget B are determined beforehand. Other parameters and their values used in the experimental studies are listed in Table 1.

At every simulation step of the agent-based model agents submit preferences and are allocated to schools. However, interventions are applied to the transport network in intervals. We evaluate the performance of the intervention strategies against a *null* baseline, where no interventions are being done, and against a *random* baseline, where interventions are performed randomly.

References

1. Abbasi, S., Ko, J., Min, J.: Measuring destination-based segregation through mobility patterns: application of transport card data. J. Transp. Geogr. **92**, 103025 (2021). https://doi.org/10.1016/j.jtrangeo.2021.103025
2. Abdulkadiroğlu, A., Sönmez, T.: Random serial dictatorship and the core from random endowments in house allocation problems. Econometrica **66**(3), 689–701 (1998)
3. Boterman, W.R.: Socio-spatial strategies of school selection in a free parental choice context. Trans. Inst. Br. Geogr. **46**(4), 882–899 (2021). https://doi.org/10.1111/tran.12454

4. Chen, D., Lü, L., Shang, M.S., Zhang, Y.C., Zhou, T.: Identifying influential nodes in complex networks. Phys. A **391**(4), 1777–1787 (2012)
5. Crescenzi, P., D'angelo, G., Severini, L., Velaj, Y.: Greedily improving our own closeness centrality in a network. ACM Trans. Knowl. Discov. Data **11**(1), 9:1–9:32 (2016). https://doi.org/10.1145/2953882
6. Dignum, E., Athieniti, E., Boterman, W., Flache, A., Lees, M.: Mechanisms for increased school segregation relative to residential segregation: a model-based analysis. Comput. Environ. Urban Syst. **93**, 101772 (2022). https://doi.org/10.1016/j.compenvurbsys.2022.101772
7. Duncan, O.D., Duncan, B.: A methodological analysis of segregation indexes. Am. Sociol. Rev. **20**(2), 210–217 (1955). https://doi.org/10.2307/2088328
8. Erdil, A., Ergin, H.: What's the matter with tie-breaking? Improving efficiency in school choice. Am. Econ. Rev. **98**(3), 669–689 (2008)
9. Holland, P.W., Laskey, K.B., Leinhardt, S.: Stochastic blockmodels: first steps. Soc. Netw. **5**(2), 109–137 (1983). https://doi.org/10.1016/0378-8733(83)90021-7
10. Ramachandran, G.S., Brugere, I., Varshney, L.R., Xiong, C.: GAEA: Graph Augmentation for Equitable Access via Reinforcement Learning. arXiv:2012.03900 (2021). arXiv: 2012.03900
11. Schelling, T.C.: Dynamic models of segregation. J. Math. Sociol. **1**(2), 143–186 (1971). https://doi.org/10.1080/0022250X.1971.9989794
12. Sissing, S., Boterman, W.R.: Maintaining the legitimacy of school choice in the segregated schooling environment of Amsterdam. Comp. Educ. **59**(1), 118–135 (2023). https://doi.org/10.1080/03050068.2022.2094580
13. Sousa, S., Nicosia, V.: Quantifying ethnic segregation in cities through random walks. Nat. Commun. **13**(1), 5809 (2022). https://doi.org/10.1038/s41467-022-33344-3
14. Stoica, V.I., Flache, A.: From schelling to schools: a comparison of a model of residential segregation with a model of school segregation. J. Artif. Soc. Soc. Simul. **17**(1), 5 (2014)
15. Zuccotti, C.V., Lorenz, J., Paolillo, R., Rodríguez Sánchez, A., Serka, S.: Exploring the dynamics of neighbourhood ethnic segregation with agent-based modelling: an empirical application to Bradford, UK. J. Ethn. Migr. Stud. **49**(2), 554–575 (2023). https://doi.org/10.1080/1369183X.2022.2100554

Author Index

F. Amigoni and A. Sinha (Eds.): AAMAS 2023 Workshops, LNAI 14456, p. 229, 2024.
https://doi.org/10.1007/978-3-031-56255-6

Printed in the United States
by Baker & Taylor Publisher Services